CONTENT on Demand

Precalculus 2nd Edition Student Resource

Practice Problem Worksheets

Cristina Berisso,
Shasta College

First published in the United States of America in 2010 by Cognella, a division of University Readers, Inc.

15 14 13 12 11 1 2 3 4 5

Printed in the United States of America

ISBN: 978-1-60927-883-0

www.cognella.com 800.200.3908

888.517.0188 www.CoDcourses.com

CONTENT on Demand

Precalculus 2nd Edition Practice Problem Worksheets

Table of Contents

CONTENT on Demand

Precalculus 2nd Edition

Practice Problem Worksheets

Review Chapter: Fundamentals of Algebra

Review Chapter Section 1: The Real Number System

Practice Problems

Check the set(s) to which each number belongs.

	Number	Natural Numbers	Integers	Rational Numbers	Irrational Numbers	Real Numbers
1.	-6					
2.	$\sqrt{16}$					
3.	0					
4.	$\frac{3}{2}$					
5.	$\sqrt{6}$					
6.	$-\frac{34}{17}$					
7.	$-\sqrt{25}$					
8.	$\frac{9}{3}$					
9.	$\frac{2\pi}{8\pi}$					
10.	$\frac{\pi}{3}$					

*Fill in each box with the appropriate symbol ($<$, $>$ or $=$) to make a **true** statement.*

11. $-1 \square 3$ ____

12. $1 \square \sqrt{3}$ ____

13. $\frac{7}{9} \square \frac{21}{27}$ ____

14. $-\frac{6}{7} \square -\frac{11}{14}$ ____

15. $-3 \square -4$ ____

16. $\frac{2}{5} \square 1$ ____

17. $\sqrt{7} \square \frac{13}{5}$ ____

18. $-\sqrt{21} \square -\frac{9}{2}$ ____

19. $\sqrt{2} \square 1$ ____

20. $\frac{3}{5} \square \frac{2}{3}$ ____

21. $\frac{5}{8} \square 0.625$ ____

22. $-2.4 \square -\sqrt{5.76}$ ____

Express the following inequalities as intervals in interval notation, and graph them.

23. $6 < x < 8$ ____________________

24. $-10 \leq x \leq -2$ ____________________

25. $-2 < x \leq 5$ ______________________________

26. $-3 < x$ ______________________________

27. $0 \leq x < 4$ ______________________________

28. $x \geq -5$ ______________________________

Express the following intervals as inequalities.

29. $(-5, 8]$ _______________

30. $(-9, -3)$ _______________

31. $[7, 12)$ _______________

32. $[4, \infty)$ _______________

33. $(-\infty, -2)$ _______________

34. $[-1, 0]$ _______________

Evaluate the following absolute values.

35. $|-12|$ _________

36. $|-1+6|$ _________

37. $|4-16|$ _________

38. $|-15+21|$ _________

39. $\left|\frac{3}{2} - \frac{17}{2}\right|$ _________

Find the distance between the following numbers on the number line.

40. 0 and -4 __________

41. 3 and -1 __________

42. 12 and 14 __________

43. -7 and -12 __________

44. -3 and 16 __________

45. $-\pi$ and 2π __________

Review Chapter Section 2: Algebraic Operations

Practice Problems

Write the following numbers in scientific notation.

1. 1,450.32 ____________

2. 2,001,000,000 ____________

3. 45,020 ____________

4. 0.000352 ____________

5. 0.000000000407 ____________

6. 0.1005 ____________

Decide whether each term below is a binomial.

Binomial?

7. $3x^2y - 7y^3$ ________

8. $12x^2 - 8xy + 7$ ________

9. $35x^3z^5y$ ________

10. $-3y + 2x^2z^3y$ ________

11. $-5x + 6x^3z^2$ ________

12. $5x - 2y + 1$ ________

13. x^2y^5z ________

14. $1 - 4x^4z$ ________

Simplify the following by removing parentheses and combining like terms.

15. $x + 3 + 2(x + 4) - 3(2x + 3)$ ________________

16. $-2x - 3(2 - 4x) + 4(3x - 7)$ ________________

17. $2x - [3x - (1 - 2x)]$ ________________

18. $7x - 3[4 - 2(1 - x)]$ ________________

19. $-2x - 3[7x - [6 - 4(3 - 2x)] + 2]$ ________________

Perform the indicated operations and simplify by combining like terms.

20. $(x-2)(x^2-x-1)$ ________________

21. $(2x^3+5xy^2)^2$ ________________

22. $(ax^2-by)(ax^2+by)$ ________________

23. $(3x^2-4y)^2$ ________________

24. $(2x-3y)(4x^2+6xy+9y^2)$ ________________

Factor the greatest common factor from the following expressions.

25. $6x^2y^3-18x^5y^2z+3x^2y^2$ ________________

26. $24x^3y^3z^9-16x^4y^{10}z^5+12x^3y^3z^5$ ________________

27. $27\,x^2\,z^3+45\,x\,z^5-72\,x^3\,z^2$ ________________

28. $-21\,y^5\,x^4+14\,y^2\,x^3-42\,y\,x^5$ ________________

29. $-70\,y^4\,z^5\,x^3+105\,y^3\,z^2\,x^4-35\,z^4\,x^6$ ________________

Factor the following expressions by grouping.

30. $6xy+4y-21x-14$ ________________

31. $4xz-10xy+6wz-15wy$ ________________

32. $4x^2y+16xyz+3xz+12z^2$ ________________

33. $20x^2yz^2+8xz^3-15xy^2-6yz$ ________________

34. $10xyz-2x^2y^2-15z^2+3xyz$ ________________

Factor the following binomials.

35. $4x^2-25$ ________________

36. $144x^2-9y^2$ ________________

37. $\dfrac{x^2}{9}-16y^4$ ________________

38. $27y^3-1$ ________________

39. $8x^3+125y^6$ ________________

Factor the following trinomials.

40. $x^2 - 6x + 8$ ____________________

41. $x^2 + 9x + 14$ ____________________

42. $x^2 - 2x - 3$ ____________________

43. $4x^2 - 17x + 15$ ____________________

44. $6x^2 + 23x - 4$ ____________________

45. $8x^2 + 11x + 3$ ____________________

Review Chapter Section 3: Linear Equations and Inequalities

Practice Problems

Solve for x.

1. $3x - 7 = 5x + 6$ __________

2. $4x - 2(x - 5) = -3x - 8$ __________

3. $2(2x - 1) - 5(2 + x) = 10$ __________

4. $4(x - 1) + 8 + 5x = 3(x - 2)$ __________

5. $2(x - 14) - 5(13 + 2x) = 39$ __________

6. $\frac{1}{2}(3x - 5) = 6 - 2x$ __________

7. $\frac{1}{4}(x - 6) + \frac{3}{2} = -2x + \frac{1}{2}$ __________

8. $\frac{1}{10}(x - 7) = \frac{1}{18}\left(\frac{x}{5} + 5\right)$ __________

9. $3(4 - 2x) + 4(3x - 6) = -2[3x - 5(1 - 4x)]$ __________

10. $4 - [3x - 2(5 - 4x)] - (3 - 2x) = 2[6x - 4(2x - 1)]$ __________

Solve the following linear inequalities and write your answer in interval notation.

11. $2(x + 5) > 3x - 1$ __________

12. $4(3 - 2x) \leq -4$ __________

13. $3(x - 2) \geq 4(2 + x)$ __________

14. $-3x - 2(3x - 4) > 6 - 5(2x - 4)$ __________

15. $2x - 4(5 - 3x) \geq -7 - 5(2 - 3x) + 4x$ __________

16. $\frac{3x + 7}{5} < 1$ __________

17. $\frac{1}{4}(x - 5) \leq x - 3$ __________

18. $\frac{2}{3}(x + 2) \geq \frac{1}{5}(2x + 3)$ __________

Solve each system of equations by ***substitution****.*

19. $\begin{Bmatrix} x + y = 4 \\ 2x + 3y = 15 \end{Bmatrix}$ ____________________

20. $\begin{Bmatrix} 3x + y = 5 \\ 4x + 2y = 2 \end{Bmatrix}$ ____________________

21. $\begin{Bmatrix} y + 3x = 2 \\ 4x - 2y = 6 \end{Bmatrix}$ ____________________

22. $\begin{Bmatrix} x - 3y = -10 \\ -5x + 2y = 24 \end{Bmatrix}$ ____________________

23. $\begin{Bmatrix} -x + 4y = -21 \\ -2x + 3y = -12 \end{Bmatrix}$ ____________________

Solve each system of equations by ***elimination****.*

24. $\begin{Bmatrix} 3x + 2y = -2 \\ -2x - 4y = 0 \end{Bmatrix}$ ____________________________

25. $\begin{Bmatrix} 2x + 3y = 8 \\ 3x + 5y = 11 \end{Bmatrix}$ ____________________________

26. $\begin{Bmatrix} 5x - 3y = 16 \\ 2x + 6y = 4 \end{Bmatrix}$ ____________________________

27. $\begin{Bmatrix} 4x - 3y = 15 \\ -7x + 2y = -10 \end{Bmatrix}$ ____________________________

28. $\begin{Bmatrix} -4x + 2y = -7 \\ 8x - 5y = 18 \end{Bmatrix}$ ____________________________

Match each of the following systems of equations with its appropriate answer pair. Write the letter of the correct answer in the box next to each system. Use the method of your choice to solve the systems.

29. $\begin{Bmatrix} 4x - 7y = -1 \\ -2x + 5y = -1 \end{Bmatrix}$ ☐ **A.** $(2,1)$

30. $\begin{Bmatrix} -x + 3y = -5 \\ 2x + 7y = -16 \end{Bmatrix}$ ☐ **B.** $(-1,2)$

31. $\begin{Bmatrix} -5x + 2y = -12 \\ 3x - 5y = 11 \end{Bmatrix}$ ☐ **C.** $(1,-2)$

32. $\begin{Bmatrix} 2x + 2y = 6 \\ -x + 9y = 17 \end{Bmatrix}$ ☐ **D.** $(1,2)$

33. $\begin{Bmatrix} 4x - 2y = -8 \\ -x - 6y = -11 \end{Bmatrix}$ ☐ **E.** $(-2,1)$

34. $\begin{Bmatrix} x + 2y = 0 \\ 2x + 3y = -1 \end{Bmatrix}$ ☐ **F.** $(-1,-2)$

35. $\begin{Bmatrix} 2x + y = 5 \\ 3x - y = 5 \end{Bmatrix}$ ☐ **G.** $(2,-1)$

36. $\begin{Bmatrix} 2x - 3y = 8 \\ 4x + 2y = 0 \end{Bmatrix}$ ☐ **H.** $(-2,-1)$

Review Chapter Section 4: Rational Expressions

Practice Problems

Simplify.

1. $\dfrac{2xy}{x^2y^2 - xy}$ ____________________

2. $\dfrac{3x^2 + 2xy}{3xy + 2y^2}$ ____________________

3. $\dfrac{2x^3 + 8x^2y}{3x^2y + 12xy^2}$ ____________________

4. $\dfrac{6x^2 - 3xy}{4x^2 - y^2}$ ____________________

5. $\dfrac{4x^3y - 36xy^3}{8x^2y^2 + 24xy^3}$ ____________________

Find an equivalent rational expression with denominator $6x^2 - x - 2$.

6. $\dfrac{5x+1}{3x-2}$ ____________________

7. $\dfrac{x-7}{2x+1}$ ____________________

8. $\dfrac{3x-2}{2x+1}$ ____________________

Find an equivalent rational expression with denominator $4x^2 + 4x - 15$.

9. $\dfrac{3x-4}{2x+5}$ ____________________

10. $\dfrac{5x+2}{2x-3}$ ____________________

11. $\dfrac{2x-3}{2x+5}$ ____________________

Simplify.

12. $\dfrac{2x^2 - 3x - 20}{2x + 5}$ ____________

13. $\dfrac{10x^2 - 3x - 4}{2x + 1}$ ____________

14. $\dfrac{12x^2 + 20x + 3}{4x^2 - 9}$ ____________

15. $\dfrac{10x^2 - 29x + 21}{4x^2 - 4x - 3}$ ____________

16. $\dfrac{18x^2 - 2}{12x^2 - x - 1}$ ____________

Multiply and simplify.

17. $\dfrac{2x + 10}{x^2 + x - 56} \cdot \dfrac{x + 8}{3x + 15}$ ____________

18. $\dfrac{7x - 56}{-4x + 32} \cdot \dfrac{x^2 + 10x + 9}{x + 9}$ ____________

19. $\dfrac{16x^2 - 1}{8x + 4} \cdot \dfrac{12x + 28}{12x^2 + 31x + 7}$ ____________

20. $\dfrac{4x^2 - 5x - 9}{16x - 12} \cdot \dfrac{32x - 24}{4x^2 - 9x}$ ____________

21. $\dfrac{2x^2 + 9x - 5}{3x^2 + 17x + 10} \cdot \dfrac{3x^2 - 13x - 10}{2x^2 + 15x - 8}$ ____________

Divide and simplify.

22. $\dfrac{33x^3 - 22x^2}{15x - 10} \div \dfrac{121x^2}{25x - 100}$ ____________

23. $\dfrac{5x^2 + 9x + 4}{5x + 4} \div \dfrac{x^2 + 3x + 2}{x - 1}$ ____________

24. $\dfrac{6x^2 + 13x + 5}{15x + 25} \div \dfrac{4x^2 - 1}{20x + 5}$ ____________

25. $\dfrac{8x^2 - 38x + 9}{12x - 9} \div \dfrac{4x^2 - x}{48x - 36}$ ____________

26. $\dfrac{4x^2 + 4x + 1}{4x^2 - x - 14} \div \dfrac{6x^2 + x - 1}{3x^2 - 7x + 2}$ ____________

Add or subtract as indicated and simplify if possible.

27. $\dfrac{5}{2x+5}+\dfrac{7}{4x-1}$ ____________________

28. $\dfrac{-6}{3x-7}+\dfrac{3x+1}{7-3x}$ ____________________

29. $\dfrac{2}{2x-1}+\dfrac{7}{8x^2-6x+1}$ ____________________

30. $\dfrac{2}{3x-5}+\dfrac{2x-12}{6x^2-7x-5}$ ____________________

31. $\dfrac{2}{2x+1}+\dfrac{4x-7}{2x^2+11x+5}$ ____________________

32. $\dfrac{3x-1}{2x^2-11x+15}-\dfrac{5}{x^2+x-12}$ ____________________

33. $\dfrac{3x+3}{3x^2+5x+2}+\dfrac{3x+7}{3x^2-x-2}$ ____________________

34. $\dfrac{5x-5}{2x^2+5x+2}-\dfrac{2x-1}{2x^2+3x-2}$ ____________________

Perform the following operations; simplify if possible. Match each item with the correct answer by writing the letter of the correct answer in the box next to each item.

35. $\left(2-\dfrac{1}{x}\right)\div\left(4-\dfrac{1}{x^2}\right)$ ☐

36. $\left(\dfrac{1}{2x}\right)\div\left(\dfrac{2}{x}+\dfrac{1}{y}\right)$ ☐

37. $\left(\dfrac{2}{3}-\dfrac{y}{x}\right)\div\left(\dfrac{x}{3}-\dfrac{y}{2}\right)$ ☐

38. $\left(\dfrac{3}{xy}\right)\div\left(\dfrac{6}{x}-\dfrac{6}{y}\right)$ ☐

39. $\left(\dfrac{4}{x^2}\right)\div\left(\dfrac{2}{x}+\dfrac{2}{y}\right)$ ☐

A. $\dfrac{2}{x}$

B. $\dfrac{2y}{x(x+y)}$

C. $\dfrac{1}{2(y-x)}$

D. $\dfrac{y}{2(x+2y)}$

E. $\dfrac{x}{2x+1}$

Solve for x.

40. $1+\frac{3}{x-2}=\frac{2x+7}{x-2}$ ______________

41. $\frac{x-2}{3}=\frac{3x-1}{7}$ ______________

42. $\frac{x+1}{x-1}=\frac{x+2}{x+6}$ ______________

43. $\frac{2x}{5x-2}-\frac{2}{3}=\frac{1}{5x-2}$ ______________

Review Chapter Section 5: Radical Expressions

Practice Problems

Simplify; express your answers with positive exponents. Assume that all variables represent positive numbers.

1. $\dfrac{\sqrt[5]{x^{-3}}}{\sqrt[5]{x^{2}}}$ ________________

2. $\sqrt{\dfrac{a^{2}x^{-6}}{(3x)^{-2}}}$ ________________

3. $\dfrac{\sqrt{xy^{3}}}{\sqrt{9x^{-1}y^{3}}}$ ________________

4. $\sqrt{18x^{3}yz}\cdot\sqrt{72xy^{5}z^{4}}$ ________________

5. $\dfrac{\sqrt[3]{108x^{7}y^{9}}}{\sqrt[3]{2xy^{2}}}$ ________________

Perform the indicated operations and combine. Assume that all variables represent positive numbers.

6. $\left(5\sqrt{x}-3\right)4\sqrt{x}$ ________________

7. $\left(2\sqrt{5}-5\sqrt{2}\right)^{2}$ ________________

8. $\left(4+7\sqrt{x}\right)3\sqrt{x}$ ________________

9. $\left(3\sqrt{x}-4\sqrt{y}\right)\left(2\sqrt{x}+5\sqrt{y}\right)$ ________________

10. $\left(4\sqrt{xy}-3\sqrt{z}\right)\left(4\sqrt{xy}+3\sqrt{z}\right)$ ________________

Simplify and combine like terms.

11. $\sqrt{81x} + 2\sqrt{x} - \sqrt{25x}$ ____________________

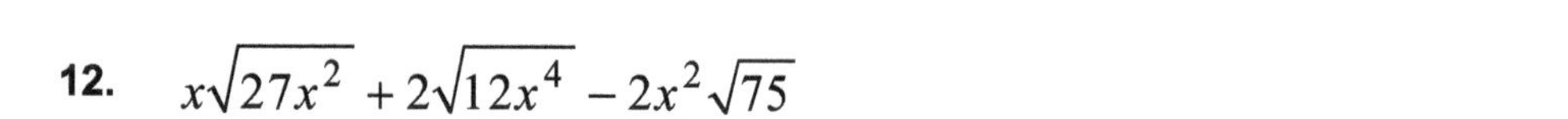

12. $x\sqrt{27x^2} + 2\sqrt{12x^4} - 2x^2\sqrt{75}$ ____________________

13. $5\sqrt[3]{40x^4} - 4x\sqrt[3]{135x}$ ____________________

14. $\sqrt[3]{\dfrac{8x^5}{27}} - 2x\sqrt[3]{\dfrac{64x^2}{125}}$ ____________________

15. $4x\sqrt{\dfrac{24x^3}{25}} - 2x\sqrt{\dfrac{6x^2}{49}} + \dfrac{\sqrt{6x^5}}{3}$ ____________________

16. Find the missing side in the right triangle below. Give the exact answer.

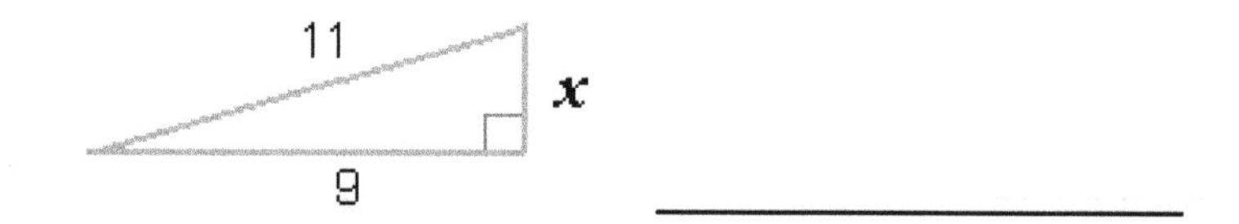

____

17. Find the missing side in the right triangle below. Give the exact answer.

____

18. Find the missing side in the right triangle below. Give the exact answer.

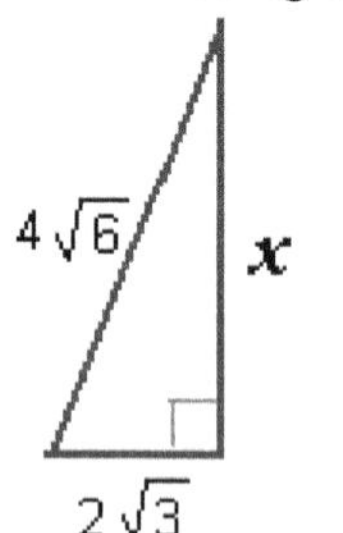

____

Solve the following equations. Match each equation with the correct answer by writing the letter of the correct answer in the box next to each equation.

19. $\sqrt{x+9}+11=x$ ☐ **A.** $x=22$

20. $\sqrt{34+4x}=\sqrt{22}$ ☐ **B.** $x=9$

21. $\sqrt{3x+4}=7$ ☐ **C.** $x=10$

22. $8\sqrt{x+3}=40$ ☐ **D.** $x=-3$

23. $\sqrt{5x+7}-\sqrt{3x+11}=0$ ☐ **E.** $x=1$

24. $\sqrt{x+6}+\sqrt{x-1}=7$ ☐ **F.** $x=15$

25. $\sqrt{3x-2}+\sqrt{3x-11}=9$ ☐ **G.** $x=7$ and $x=16$

26. $3\sqrt{2x}=2\sqrt{x+5}$ ☐ **H.** $x=2$

27. $\sqrt{4x+5}-\sqrt{3x-12}=0$ ☐ **I.** No solution

28. $\sqrt{5x+4}-\sqrt{5x-4}=2$ ☐ **J.** $x=\frac{10}{7}$

Perform the following operations.

29. $(3+2i)+(-2+7i)$ ______________

30. $(5-3i)+(3+7i)$ ______________

31. $(-3+2i)-(4-i)$ ______________

32. $(2-5i)-(-2-3i)$ ______________

33. $(3-3i)\cdot(2+5i)$ ______________

34. $(2+4i)\cdot(-3+2i)$ ______________

35. $\frac{2+5i}{1-i}$ ______________

36. $\frac{3-7i}{2+3i}$ ______________

Review Chapter Section 6: Quadratic Equations

Practice Problems

Solve the following quadratic equations using the square root property.

1. $x^2 = 121$ ________________

2. $x^2 = \frac{25}{169}$ ________________

3. $x^2 = -64$ ________________

4. $x^2 = 125$ ________________

5. $(x-2)^2 = 49$ ________________

6. $(x+5)^2 = -36$ ________________

7. $(x+3)^2 = -45$ ________________

8. $(x+1)^2 = 3$ ________________

9. $(x-3)^2 = 5$ ________________

10. $\left(x-\frac{1}{2}\right)^2 = -\frac{9}{4}$ ________________

11. $(x+1)^2 = -\frac{7}{4}$ ________________

Find the number that must be added to each expression to make it a perfect square trinomial. Also note the perfect square in factor form.

12. $x^2 + 6x$ ____________________

13. $x^2 - 14x$ ____________________

14. $x^2 - 10x$ ____________________

15. $x^2 + x$ ____________________

16. $x^2 - \frac{9}{4}x$ ____________________

Solve the following quadratic equations by completing the square.

17. $x^2 - 5x + 1 = 0$ _______________

18. $x^2 + x - 4 = 0$ _______________

19. $x^2 - x + \frac{5}{4} = 0$ _______________

20. $x^2 + 4x + 16 = 0$ _______________

21. $x^2 - 4x - 11 = 0$ _______________

22. $9x^2 - 6x + 1 = 0$ _______________

23. $2x^2 - 6x + 5 = 0$ _______________

24. $3x^2 - 4x - 1 = 0$ _______________

Solve the following quadratic equations by using the quadratic formula.

25. $x^2 + x + 1 = 0$ ________________

26. $x^2 - 4x - 6 = 0$ ________________

27. $x^2 - x + \frac{1}{4} = 0$ ________________

28. $x^2 + 6x + 5 = 0$ ________________

29. $-x^2 + 2x + 5 = 0$ ________________

30. $-2x^2 - 7x + 3 = 0$ ________________

31. $3x^2 + 5x + 1 = 0$ ________________

32. $4x^2 - 10x + 3 = 0$ ________________

33. $9x^2 - 12x + 4 = 0$ ________________

By studying the discriminant in the quadratic formula, decide the number and type of solutions expected for the following quadratic equations.

	Equation	**One rational solution**	**Two rational solutions**	**Two irrational solutions**	**Two non-real solutions**
34.	$3x^2 - 4x + 1 = 0$				
35.	$-x^2 + 2x - 9 = 0$				
36.	$6x^2 - 7x + 5 = 0$				
37.	$2x^2 - 3x - 2 = 0$				
38.	$5x^2 - 4x - 6 = 0$				
39.	$2x^2 + x - 3 = 0$				
40.	$9x^2 + 12x + 4 = 0$				

Review Chapter Section 7: The Rectangular Coordinate System

Practice Problems

1. Mark the (x, y) coordinates of the plotted points on the graph below.

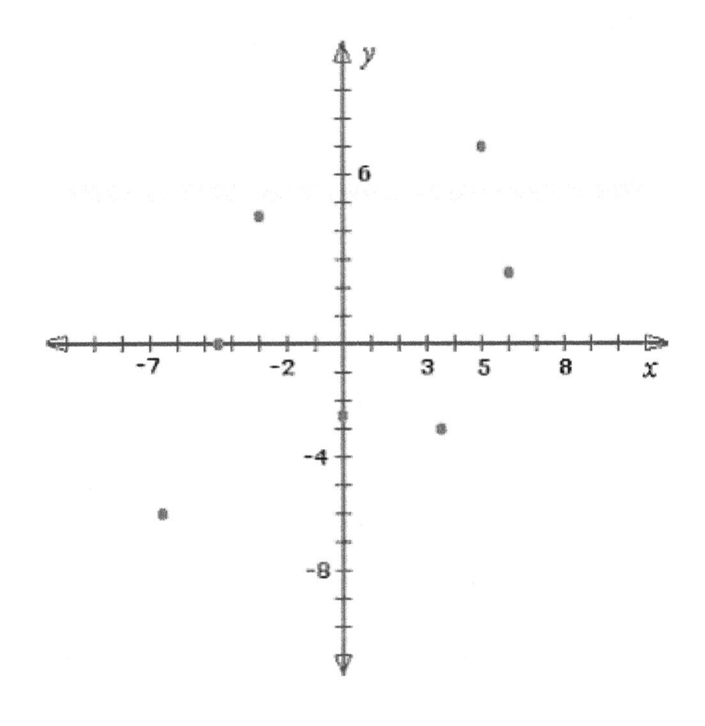

2. Plot the line described by the following equation: $2x - 7y = 5$.

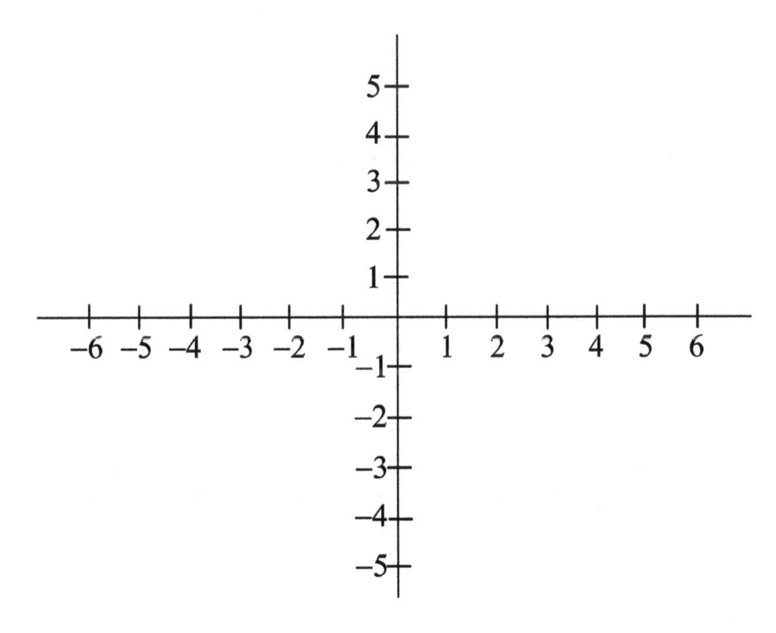

3. Plot the line described by the following equation: $3x+3y=2$.

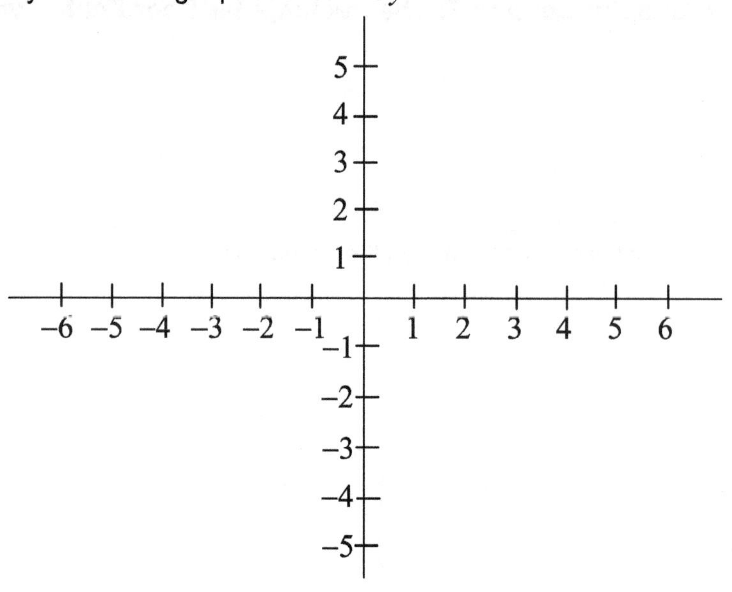

4. Plot the line described by the following equation: $-x+5y=-3$.

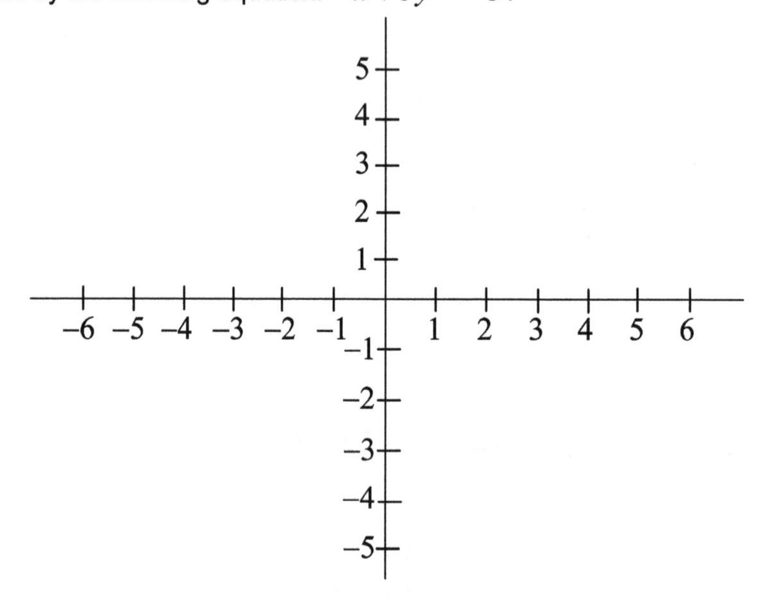

5. Plot the line described by the following equation: $-5x-2y=4$.

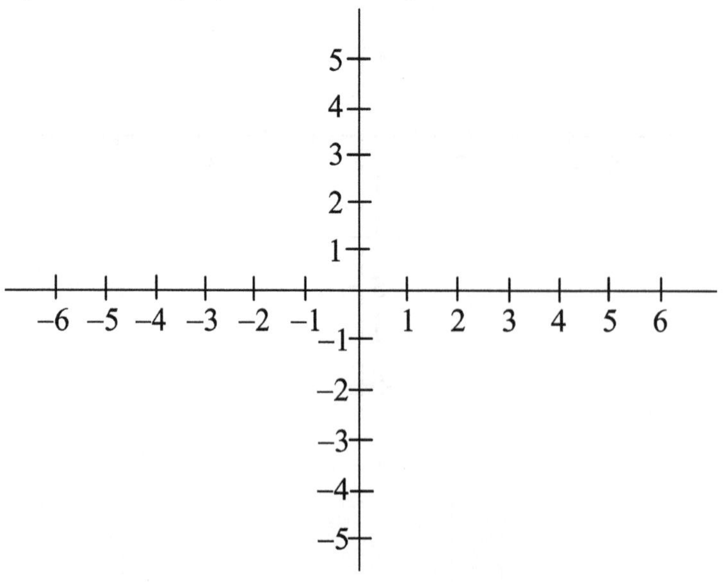

Find the distance between each given pair of points on the plane. Give the exact answer.

6. $(0,2)$ and $(0,4)$ ____________

7. $(3,0)$ and $(0,3)$ ____________

8. $(0,-4)$ and $(-4,4)$ ____________

9. $\left(\frac{5}{3},0\right)$ and $\left(\frac{5}{3},4\right)$ ____________

10. $\left(3,-\frac{2}{5}\right)$ and $\left(5,-\frac{2}{5}\right)$ ____________

11. $\left(\frac{5}{2},-4\right)$ and $\left(-\frac{1}{2},3\right)$ ____________

12. $(3,-2)$ and $(5,4)$ ____________

13. $(-1,-3)$ and $(4,-1)$ ____________

14. $\left(\sqrt{3},3\right)$ and $(0,-2)$ ____________

15. $(-3,0)$ and $\left(4,-\sqrt{5}\right)$ ____________

Write the equation of the circle with the given characteristics.

16. Center at $(0,0)$ and radius 5 ____________

17. Center at $(0,-3)$ and radius 7 ____________

18. Center at $(2,3)$ and radius 6 ____________

19. Center at $\left(\frac{1}{2},4\right)$ and radius $\frac{3}{2}$ ____________

20. Center at $\left(\frac{3}{4}, \frac{1}{3}\right)$ and radius $\frac{1}{2}$ ____________________

21. Center at $(3, -1)$ and radius 2 ____________________

22. Center at $(2, 3)$ and radius $\sqrt{5}$ ____________________

23. Center at $\left(\frac{3}{5}, -2\right)$ and radius $\sqrt{\frac{2}{3}}$ ____________________

24. Center at $(0, -5)$ and radius $2\sqrt{3}$ ____________________

25. Center at $(-1, -6)$ and radius $\frac{3}{2}$ ____________________

Review Chapter Section 8: Polynomial and Rational Inequalities

Practice Problems

Solve the following inequalities. Give your answer in interval notation.

1. $x^2 + 2x - 3 < 0$ ________________

2. $x^2 - x - 6 \geq 0$ ________________

3. $x^2 + 3x - 10 \leq 0$ ________________

4. $3x^2 - 5x - 2 > 0$ ________________

5. $-2x^2 + 19x - 35 > 0$ ________________

6. $2x^2 + 7x - 15 \leq 0$ ________________

7. $6x^2 + x - 2 > 0$ ________________

8. $6x^2 + 17x + 5 < 0$ ________________

9. $3x^2 - 14x + 8 \geq 0$ ________________

10. $x^2 + x + 1 < 0$ ________________

11. $\dfrac{1}{x-3} < 0$ ________________

12. $\dfrac{4}{x+5} \leq 2$ ________________

13. $\dfrac{13}{x+9} \geq 1$ ________________

14. $\dfrac{x-3}{x-4} \geq 0$ __________________

15. $\dfrac{-12}{x-5} \geq 2$ __________________

16. $\dfrac{x}{x-4} \geq -3$ __________________

17. $\dfrac{x-6}{x+3} \geq 0$ __________________

18. $\dfrac{3x-1}{x+2} < 1$ __________________

19. $\dfrac{5x+2}{x-1} \geq -3$ __________________

20. $\dfrac{2x-7}{x+4} \leq 5$ __________________

21. $\dfrac{3x+2}{2x-1} > 4$ __________________

22. $\dfrac{4x^2-8x-5}{2x-3} \leq 0$ __________________

CONTENT on Demand

Precalculus 2nd Edition

Practice Problem Worksheets

Chapter 1: Functions—An Overview

Chapter 1 Section 1: Introduction to Functions

Practice Problems

Find the domain and the range of the relations represented by the following arrow diagrams, and state whether the relation does or does not represent a function.

1. 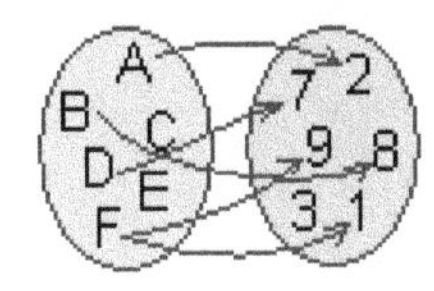

____

2. 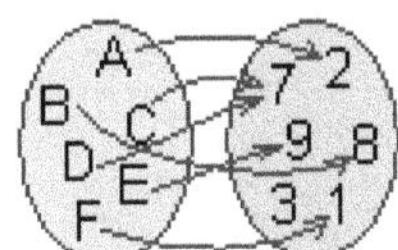

____

3. 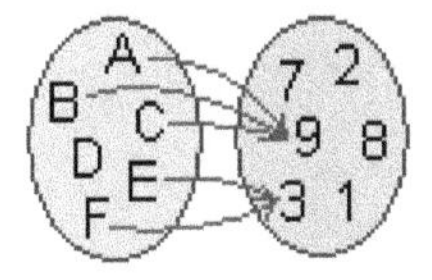

____

4. 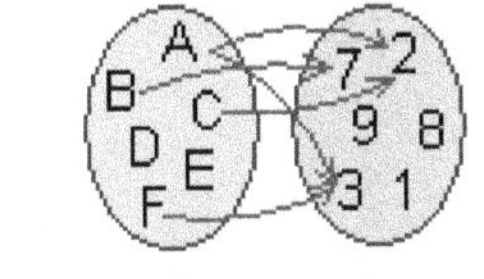

____

5. 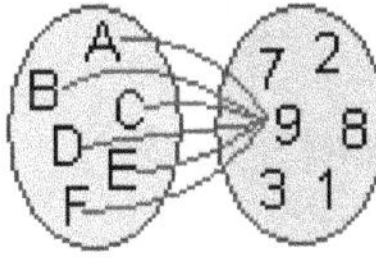

____

6. 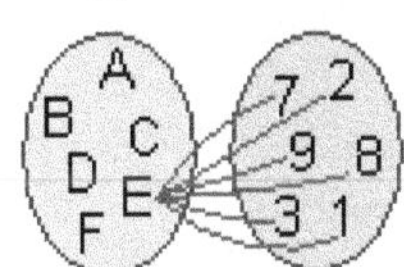

____

Find the domain and the range of the relations represented by the following input-output (x, y) pairs, and state whether the relation does or does not represent a function.

7. $\{(a,2),(b,-3),(c,5),(d,4),(b,3)\}$ ______________________________

8. $\{(a,-3),(b,-3),(c,3),(d,-3),(e,3)\}$ ______________________________

9. $\{(-2,a),(1,c),(0,d),(-1,a),(2,c)\}$ ______________________________

10. $\{(-2,1),(1,3),(0,-2),(-1,1),(-2,0)\}$ ______________________________

11. $\{(-1,1),\ (0,-1),(1,-2),(0,3),(2,5)\}$ ______________________________

Find the domain and range of each of the relations defined by the following input-output tables, and state whether the relation is or is not a function.

12.

Input	2	−2	1	−3	0
Output	5	−5	−1	2	1

____

13.

Input	1	−3	4	−2	4
Output	4	5	1	0	3

____

14.

Input	80	12	100	40	5
Output	0.1	0.1	0.1	0.1	0.1

____

15.

Input	0.1	0.2	0.3	0.1	0.2
Output	1	1	2	−1	−1

____

16.

Input	20	−12	11	−13	10
Output	3	−3	−2	1	2

____

17. *The following bar graph shows the distribution of test grades in a small class. Build a table of input-output values based on the data shown. Do the data represent a function?*

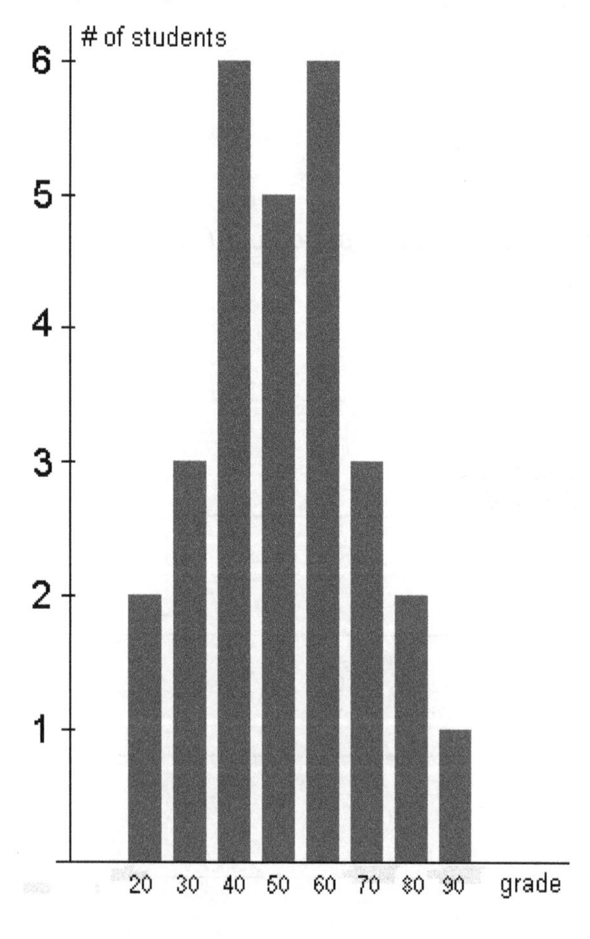

Input (grade)	**Output** (# of students)

18. *A graph of a patient's temperature over time is shown below. Complete the table of input-output values based on the information provided in the graph.*

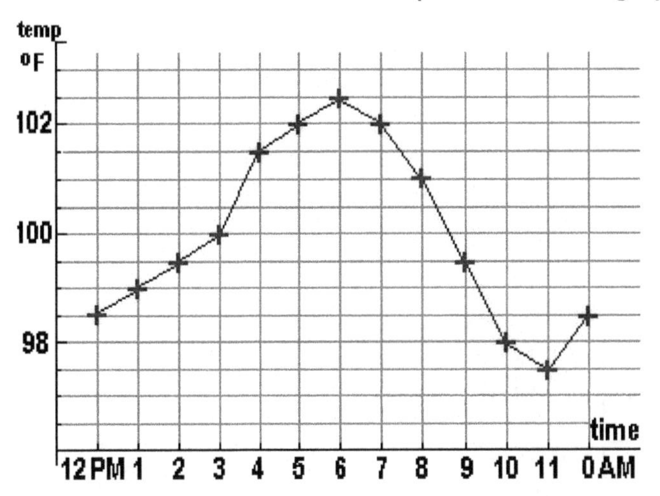

Time	Temp (F)
2 PM	
	102.5
	98
11PM	

19. *The formula that relates the distance covered as a function of hours of travel for the "shinkansen" (or bullet train) of Japan that runs from Tokyo to Nagano is given below. Based on this formula, complete the table of input-output values.*

Distance [in Km] = 252 · t [hours]

Input Time t in hours	**Output**: Distance in km
	126
	252
1.5	
	630
4	

20. *The formula that relates the temperature in degrees Fahrenheit (**F**) and Celsius (**C**) is given below. Based on this formula, complete the table of input-output values.*

$$F = \frac{9}{5}C + 32$$

Input: Temp **F**	**Output**: Temp **C**
	0
	15
86	
	45
140	
176	
	100

21. *The following bar graph shows the number of calories that a person burns per minute using the treadmill in the gym under three different settings (Settings A, B and C).*

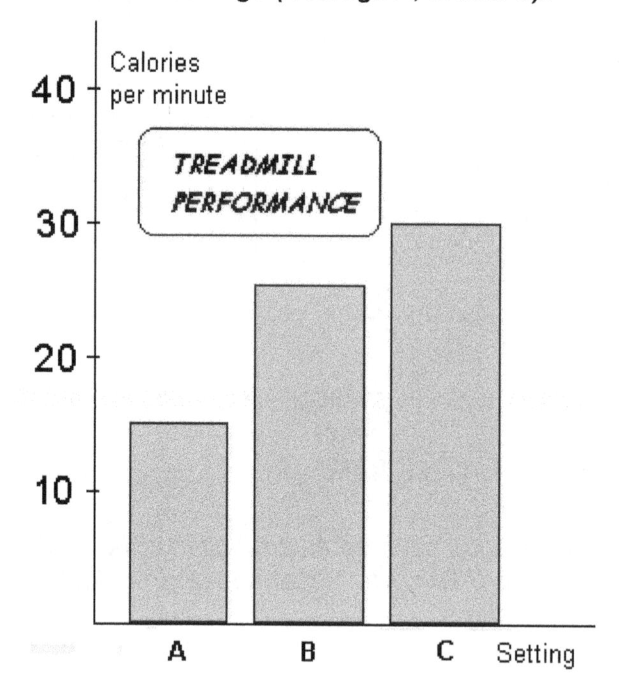

Find the following:

a. The number of calories burned by using Setting A for one half hour.

____

b. The total number of calories burned by using Setting A for 1 hour and setting B for 45 minutes.

____

c. The total number of calories burned by using Setting A for 45 minutes, Setting B for one half hour and Setting C for 15 minutes.

____

d. The total number of calories burned by using Setting A for 20 minutes, Setting B for one hour and 20 minutes and Setting C for 45 minutes.

____

e. The total number of calories burned by using Setting A for 35 minutes, Setting B for one hour and 15 minutes and Setting C for 25 minutes.

____

22. *The following table shows the number of calories for different breakfast foods.*

Food	Calories
Scrambled egg (1)	75
Toast (1)	87
Orange juice (1 cup)	90
English muffin (1)	130
Bacon (1 slice grilled)	195
Danish pastry	265
Cappuccino (medium size)	60
Corn flakes (1 serving)	110
Milk 2% (1 cup)	120
Bagel (1)	64

continued…

Use the data given in the table on page 4 to answer the following:

a. Is the relationship that uses "Calories" as input and "Food" as output a function?

____

b. Is the relationship that uses "Food" as input and "Calories" as output a function?

____

c. How many total calories are there in a breakfast that consists of 2 scrambled eggs, one bagel and one cup of orange juice?

____

d. How many total calories are there in a breakfast that consists of 2 slices of grilled bacon, one cappuccino and one-half of a Danish pastry?

____

e. How many total calories are there in a breakfast that consists of one cup of milk, two servings of corn flakes and two pieces of toast?

____

23. *Draw a point diagram of the data shown in the following table and connect the (x, y) points of consecutive input values with line segments.*

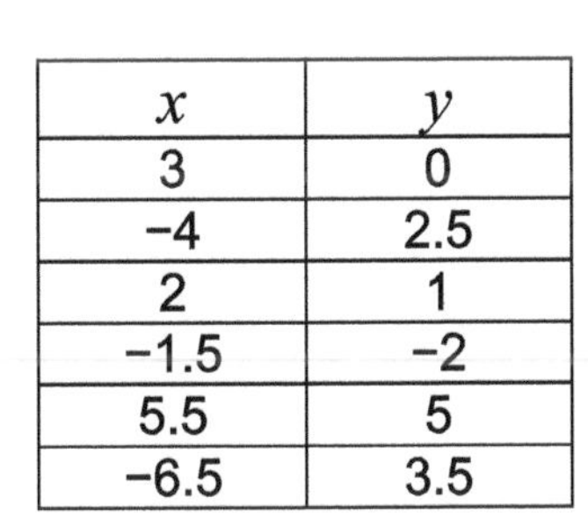

x	y
3	0
−4	2.5
2	1
−1.5	−2
5.5	5
−6.5	3.5

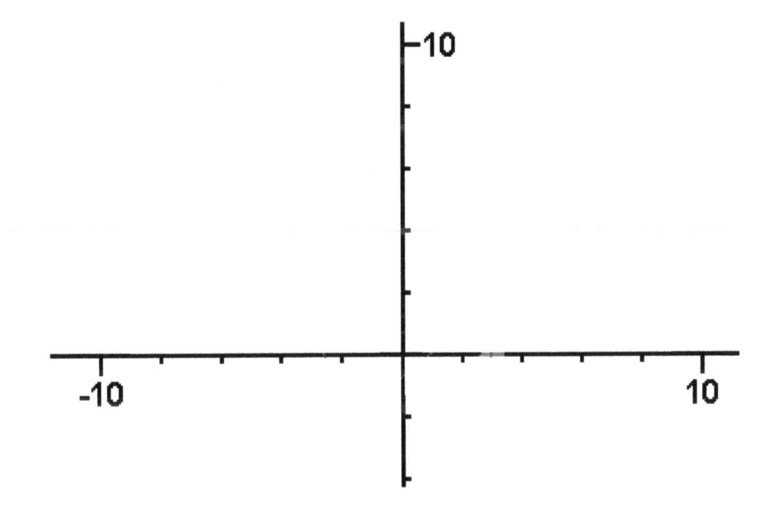

24. *Put a check mark in the box under the graphs that represent functions.*

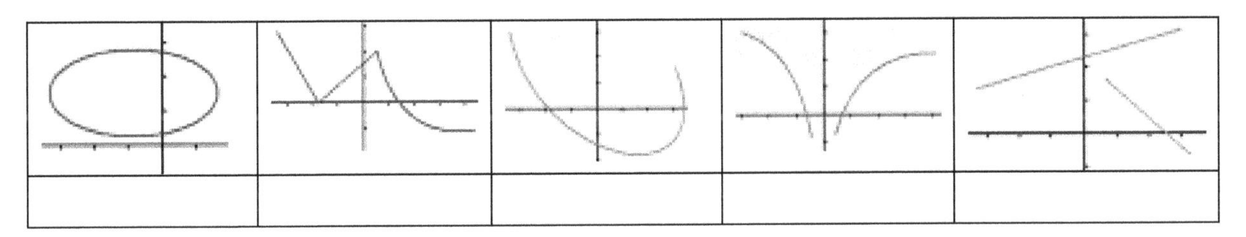

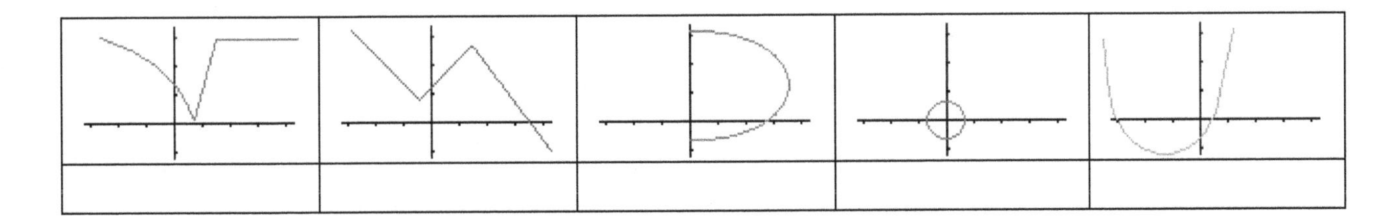

Chapter 1 Section 2: Function Notation

Practice Problems

Give a mathematical expression for each of the following functions.

1. The area (A) of the circle as a function of its radius (r).

____

2. The area (A) of the circle as a function of its diameter (d). [Hint: use the answer from #1 and the fact that $r = d/2$.]

____

3. The radius (r) of the circle as a function of its area (A).

____

4. The distance (D) covered by a car traveling at 55 miles per hour as a function of the number of hours (t) of traveling time.

____

5. The area (A) of a triangle as a function of its base (b), for a triangle whose height is $1/3$ of the length of its base.

____

6. The area of the figure called an "ellipse" is given by the product of the ellipse's two semi-axes (the major semi-axis times the minor semi-axis) times π.

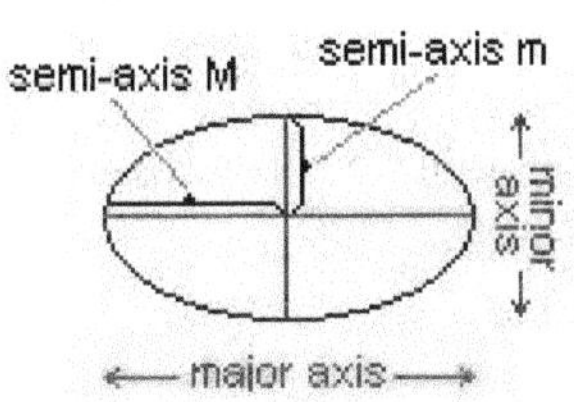

For a special case of an ellipse whose major axis is four times the length of the minor axis, find a mathematical expression for the ellipse's area (A) as a function of the major semi-axis (M).

____

7. The following expression relates the variables p and q: $p = \frac{q+2}{q-4}$. Find q as a function of p.

____

8. The following expression relates the variables p and q: $p = \frac{q-3}{q+2} - 4$. Find q as a function of p.

____

9. *The graph of a patient's temperature over time is shown below.*

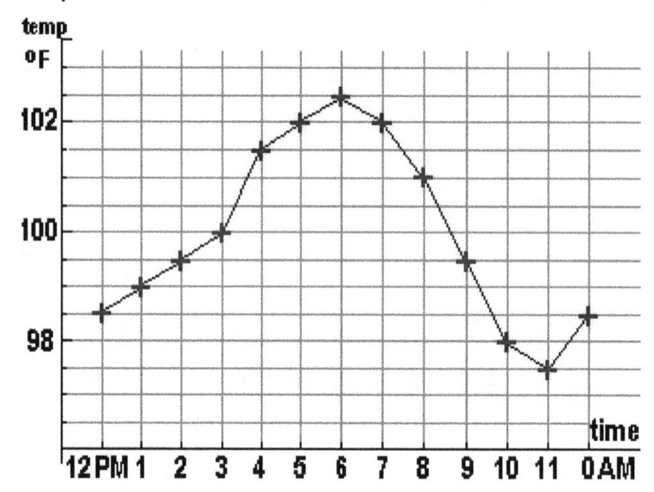

Use the graph to answer the following:

a. At what time did the patient show his maximum recorded temperature?

____

b. What was this maximum recorded temperature?

____

c. At what time did the patient show his minimum recorded temperature?

____

d. What was this minimum recorded temperature?

____

e. On two occasions, the patient's recorded temperature showed the same reading above 100° F. When were these readings recorded?

____

f. What were the two similar temperature readings above 100° F?

____

g. On two occasions, the patient's recorded temperature showed the same readings below 100° F. When were these readings recorded?

____

h. What were the two similar temperature readings below 100° F?

____

10. *The following equation defines* $f(x)$ *as a function of* x*:* $f(x) = 2x^3 - 4x + 1$.
Find the value of the function at each of the following values.

a. $x = 0$ ________________

b. $x = 1$ ________________

c. $x = \frac{3}{2}$ ________________

d. $x = -\frac{1}{2}$ ________________

e. $x = -\frac{3}{2}$ ________________

11. *The following equation defines* $f(x)$ *as a function of* x*:* $f(x) = -3x^2 + 2x$.
Find the value of the function at each of the following values.

a. $x = 0$ ________________

b. $x = 1$ ________________

c. $x = \frac{3}{2}$ ________________

d. $x = -\frac{1}{2}$ ________________

e. $x = -\frac{5}{2}$ ________________

12. *The following equation defines* $f(x)$ *as a function of* x*:* $f(x) = -5 + 2\sqrt{3 - 4x}$.
Find the value of the function at each of the following values.

a. $x = 0$ ________________

b. $x = 1$ ________________

c. $x = \frac{3}{4}$ ________________

d. $x = -\frac{5}{4}$ ________________

e. $x = \frac{1}{2}$ ________________

13. *The following equation defines* $f(x)$ *as a function of* x: $f(x) = \dfrac{-2x+1}{\sqrt{3-4x}}$.

Find the value of the function at each of the following values.

a. $x = 0$ ________________

b. $x = 1$ ________________

c. $x = \dfrac{3}{4}$ ________________

d. $x = -\dfrac{5}{4}$ ________________

e. $x = \dfrac{1}{2}$ ________________

14. *The graph of a function is given below.*

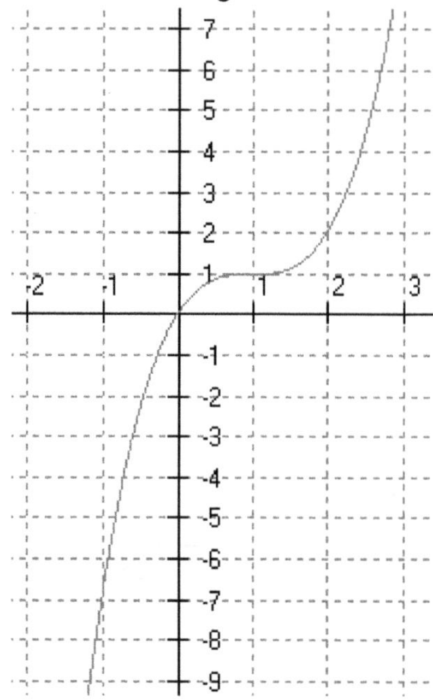

Use the graph to find the value of each of the following:

a. $f(-1)$ ________________

b. $f(0)$ ________________

c. $f(1)$ ________________

d. $f(2)$ ________________

e. $f(2) - f(-1)$ ________________

f. $3 \cdot f(1)$ ________________

g. $f(0) + f(1)$ ________________

h. $3 \cdot f(2) - 2 \cdot f(-1)$ ________________

i. $6 \cdot f(1) + 3 \cdot f(-1)$ ________________

j. $\dfrac{f(2) - 3 \cdot f(-1)}{5 \cdot f(1)}$ ________________

15. *The graph of a function is given below.*

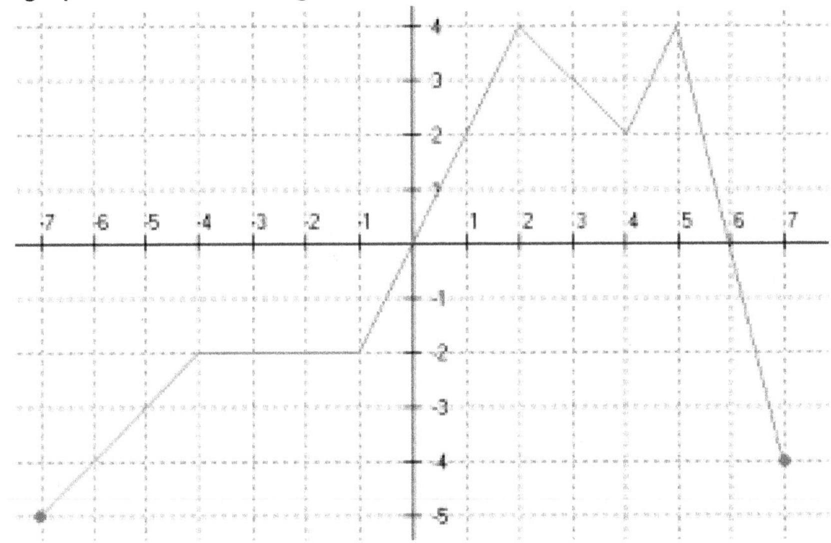

Use the graph to find the value of each of the following.

a. $f(-5)$ ________

b. $f(-3)$ ________

c. $f(1)$ ________

d. $f(6)$ ________

e. $f(4)$ ________

Use algebra to find the domain of the following functions.

16. $f(x) = \dfrac{x^2 - 3x}{4}$ ________________________

17. $f(x) = \dfrac{2x^2 + 5}{x}$ ________________________

18. $f(x) = \dfrac{2x^2}{x - 8}$ ________________________

19. $f(x) = \dfrac{3x-2}{x+5}$ ______________________________

20. $f(x) = 5 - \sqrt{3-x}$ ______________________________

21. $f(x) = \dfrac{x}{\sqrt{2-x}}$ ______________________________

22. $f(x) = \dfrac{x+4}{(x+3)^2}$ ______________________________

State the domain and range of the function defined by the graph:
Hint: Project the graph onto the x and y axes, respectively.

23.

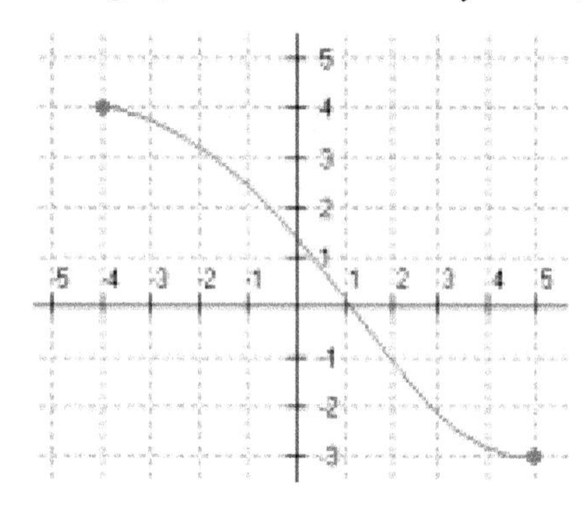

____

24.

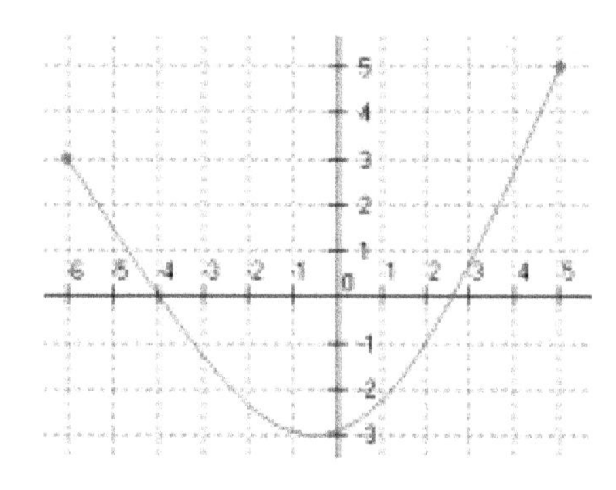

____

25.

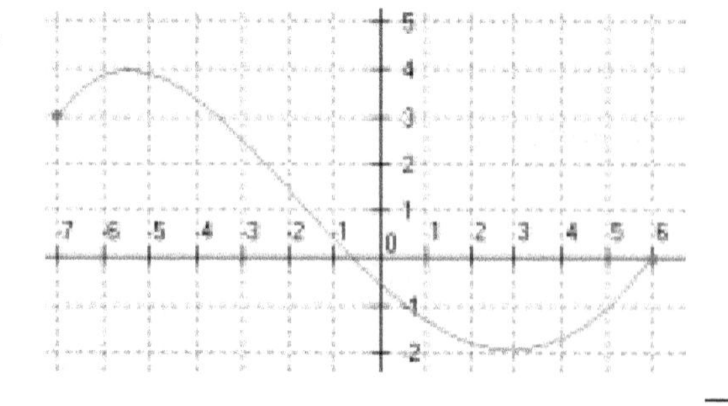

____

26.

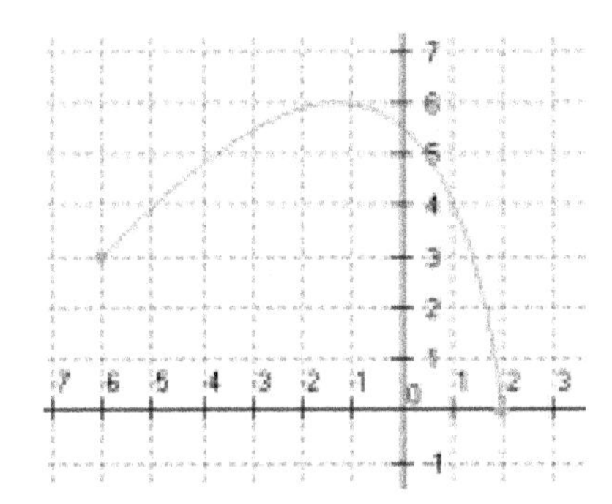

____

27.

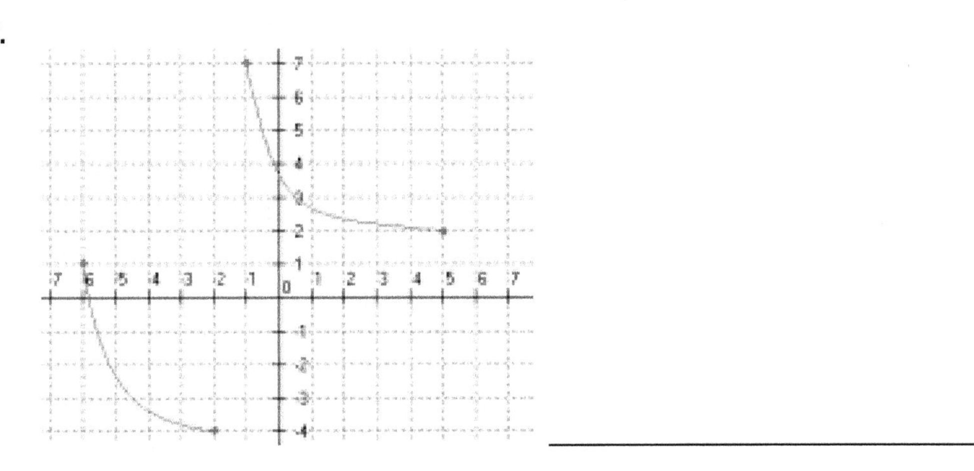

____

Chapter 1 Section 3: Functions and Their Graphs

Practice Problems

Graph each of the following functions in the space provided.

1. $f(x) = -3x + 5$

2. $f(x) = \dfrac{x - 4}{2}$

3. $f(x) = |x - 3|$

4. $f(x) = \dfrac{|x + 1|}{2} - 3$

5. $f(x) = x^2 - 3$

6. $f(x) = -x^2 + 4$

7. $f(x) = x^2 + 4x - 1$

8. $f(x) = -x^2 - 2x + 2$

9. $f(x) = 2x^2 - 6x + 5$

10. *An adult ticket for a play is $35. Children younger than 12 pay $8. No entrance is allowed to children younger than 6, and seniors (55 or older) pay $20. Draw a graph that represents the ticket price as a function of age.*

11. *A telephone company charges $0.25 for the first three minutes of an international call, and $0.10 for each additional minute. Draw a graph that represents the call charges (in cents) as a function of the duration of the call (in minutes).*

Graph each of the following piecewise functions in the space provided.

12. $$f(x)=\begin{cases} 2x-1 & \text{for } x<2 \\ -3x+4 & \text{for } x\geq 2 \end{cases}$$

13. $$f(x)=\begin{cases} x+2 & \text{for } x<0 \\ -x^2+2 & \text{for } x\geq 0 \end{cases}$$

14. $$f(x)=\begin{cases} 2x^2+3x-2 & \text{for } x\leq -1 \\ -x-1 & \text{for } x>-1 \end{cases}$$

15. $$f(x)=\begin{cases} -x^2+4 & \text{for } x<2 \\ -2 & \text{for } x=2 \\ x-5 & \text{for } x>2 \end{cases}$$

16. *Evaluate each of the following when given this piecewise function:*

$$f(x)=\begin{cases}-2x^2-5x+1 & \text{for } x<0\\ 4x-7 & \text{for } x\geq 0\end{cases}$$

[Hint: your first step is to locate the boundary point to help you decide which of the functional expressions you must use in the evaluation.]

a. $f(2)$ __________

b. $f(-1)$ __________

c. $f(0)$ __________

d. $f(-3)+f(0)$ __________

e. $f(-2)-f(-1)$ __________

f. $f(3)-f(-2)$ __________

g. $f(-1)\cdot f(5)$ __________

17. *Evaluate each of the following when given this piecewise function:*

$$f(x)=\begin{cases}-x^2+1 & \text{for } x\leq -1\\ x^2-2x & \text{for } x>-1\end{cases}$$

[Hint: your first step is to locate the boundary point to help you decide which of the functional expressions you must use in the evaluation.]

a. $f(2)$ __________

b. $f(-1)$ __________

c. $f(-3)$ __________

d. $f(-3)-f(1)$ __________

e. $f(-2)-f(-4)$ __________

f. $f(3)-f(-2)$ __________

g. $f(-2)\cdot f(5)$ __________

18. *The manager of an appliances store gives to his employees an "incentive bonus" according to their total monthly sales, as defined in the table below.*

Monthly Sales	Bonus
Less than or equal to $4000	0 (zero)
More than $4000 to $8000	4% of sales in excess of $4000
More than $8000	8% of sales in excess of $8000, plus $160

Find the "bonus" amount for each of the following cases.

a. An employee who sold goods for $7800 over the month. __________

b. An employee who sold goods for $8000 over the month. __________

c. An employee who sold goods for $8500 over the month. __________

d. An employee who sold goods for $10,500 over the month. __________

19. *The graph of a function is given below.*

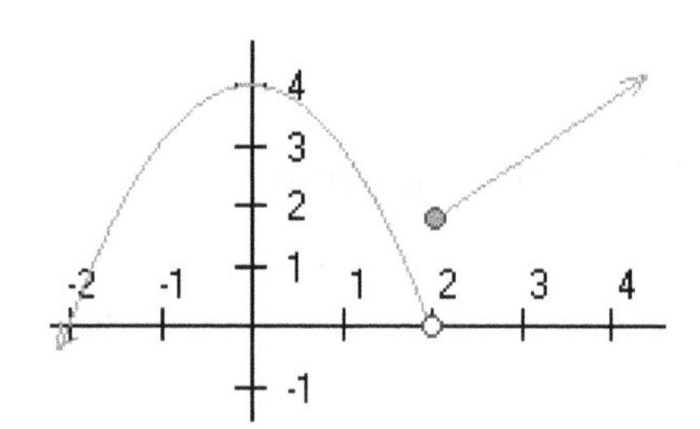

State the intervals in which the function is:

a. Increasing ________________

b. Decreasing ________________

c. Constant ________________

20. *The graph of a function is given below.*

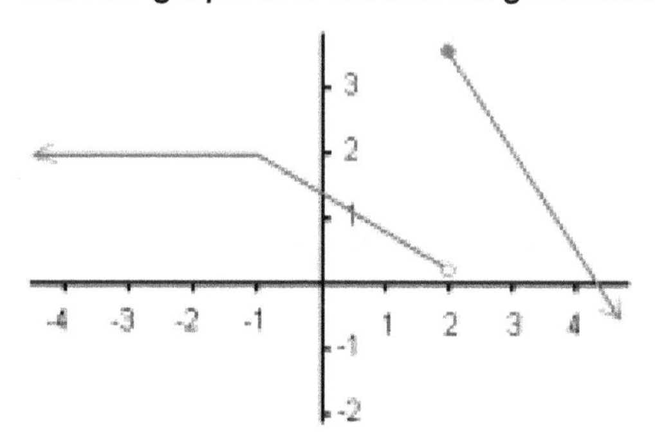

State the intervals in which the function is:

a. Increasing ________________

b. Decreasing ________________

c. Constant ________________

21. *The graph of a function is given below.*

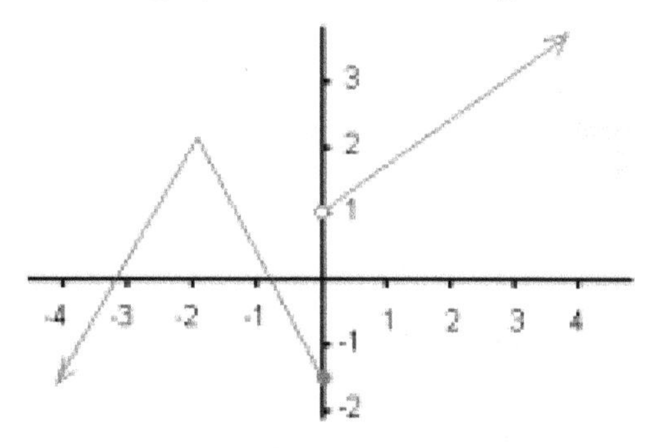

State the intervals in which the function is:

a. Increasing ________________

b. Decreasing ________________

c. Constant ________________

22. *The graph of a function is given below.*

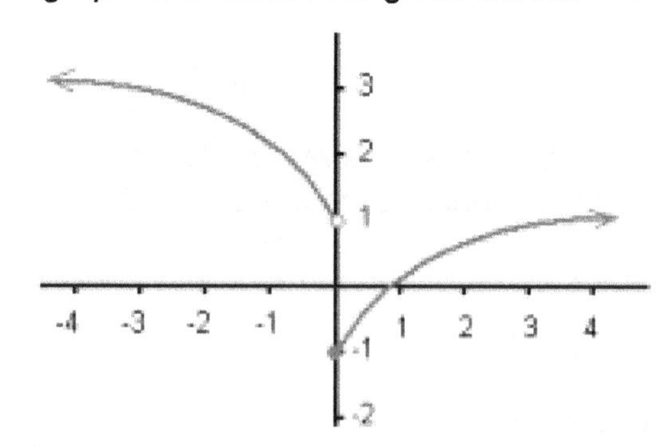

State the intervals in which the function is:

a. Increasing ________________

b. Decreasing ________________

c. Constant ________________

23. *The graph of a function is given below.*

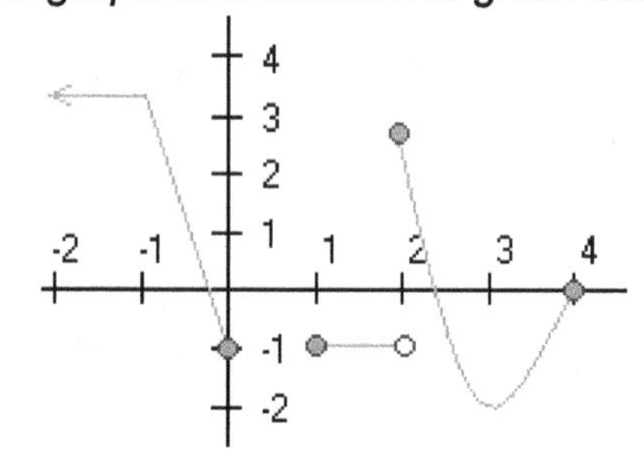

State the intervals in which the function is:

a. Increasing ________________

b. Decreasing ________________

c. Constant ________________

Match each of the following piecewise functions with the appropriate graph.

24.
$$f(x)=\begin{cases} x+3 & \text{for } -3\le x<0 \\ \dfrac{2x}{3} & \text{for } 0\le x\le 3 \end{cases}$$ ☐

A.

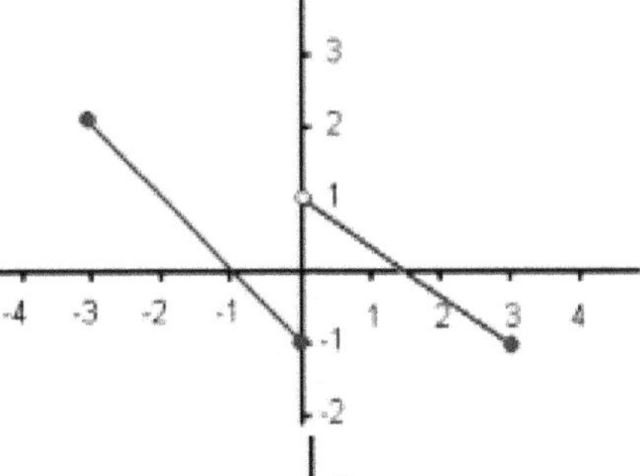

25.
$$f(x)=\begin{cases} -x-1 & \text{for } -3\le x\le 0 \\ -\dfrac{2x}{3}+1 & \text{for } 0<x\le 3 \end{cases}$$ ☐

B.

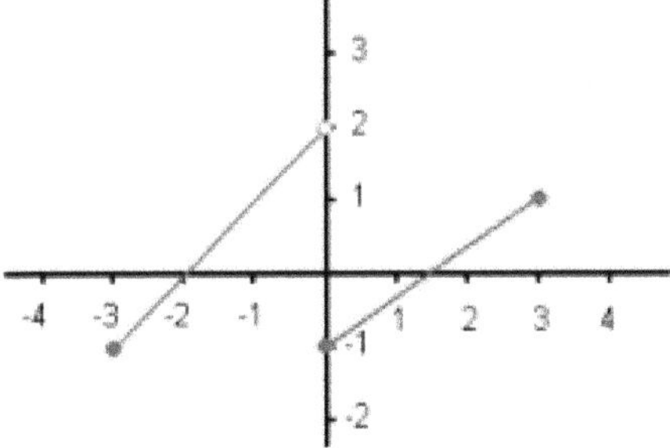

26.
$$f(x)=\begin{cases} -x-2 & \text{for } -3\le x\le 0 \\ \dfrac{3x}{2}-1 & \text{for } 0<x\le 3 \end{cases}$$ ☐

C.

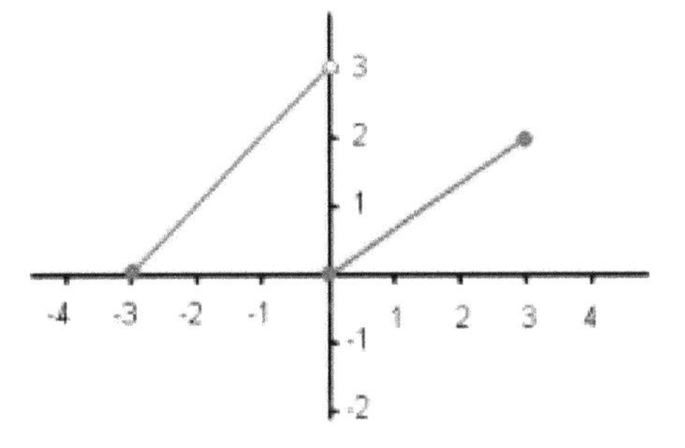

27.
$$f(x)=\begin{cases} -x+1 & \text{for } -3\le x<0 \\ \dfrac{x}{2} & \text{for } 0\le x\le 3 \end{cases}$$ ☐

D.

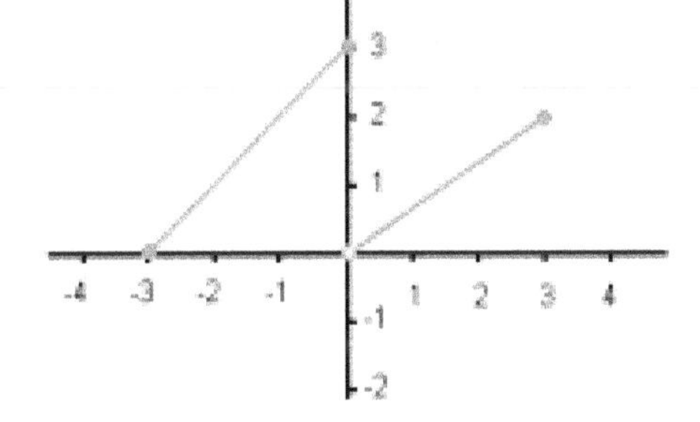

28.
$$f(x)=\begin{cases} x+3 & \text{for } -3\le x\le 0 \\ \dfrac{2x}{3} & \text{for } 0<x\le 3 \end{cases}$$ ☐

E.

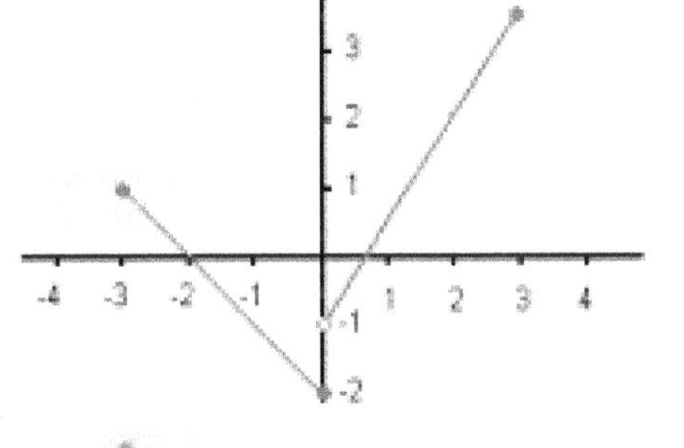

29.
$$f(x)=\begin{cases} x+2 & \text{for } -3\le x<0 \\ \dfrac{2x}{3}-1 & \text{for } 0\le x\le 3 \end{cases}$$ ☐

F.

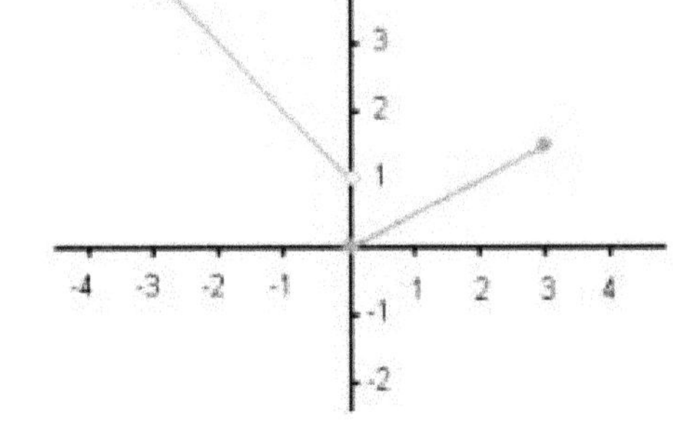

Chapter 1 Section 4: Applications of Functions

Practice Problems

1. A piece of wire x inches long is shaped like a circle. Find the expression of the circle's area as a function of the length (x) of the wire.

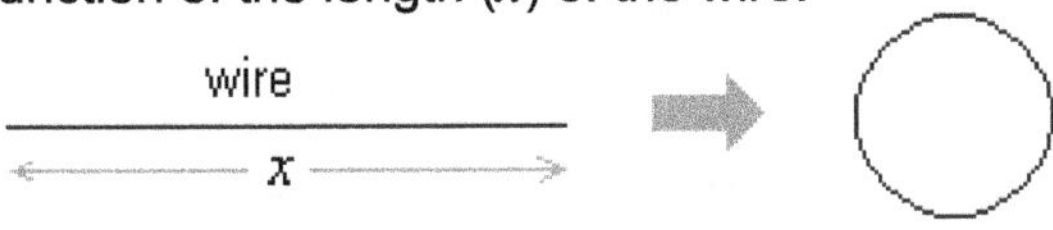

2. A piece of wire $4x$ inches long is shaped like a circle. Find the expression of the circle's area as a function of (x).

3. A piece of wire x inches long is shaped like a square. Find the expression of the square's area as a function of the length of the wire (x).

4. A piece of wire $10x$ inches long is shaped like a square. Find the expression of the square's area as a function of (x).

5. A piece of wire of length x is bent in the shape of an equilateral triangle. Find the triangle's area as a function of the wire's length (x).

6. The length of the hypotenuse of a right triangle is 9 feet. Write an expression for the triangle's perimeter as a function of one of the other sides of the triangle.

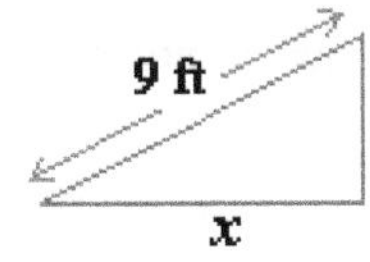

7. The length of the hypotenuse of a right triangle is 15 meters. Write an expression for the triangle's area as a function of one of the other sides of the triangle.

15 m

x

____

8. A plumber charges a fixed rate of $55 per visit, plus $48 per hour of labor. Write an expression for the plumber's total charges as a function of number of hours (x) of labor.

____

9. The total cost of producing electronic devices in a small company is given by a fixed cost of $2600 for running the plant, plus $56 per unit produced. Write an expression for the total cost for producing x units of the electronic devices.

____

10. A car rental company charges $30.40 per day for the rental of a subcompact car, plus 15 cents per mile driven.

a. Write an expression for the rental charges per day as a function of miles driven.

____

b. What is the maximum number of miles that a person renting one car for 1 day can travel if he does not want the rental bill to exceed $70?

____

c. What is the total amount that a person will be charged if he rents a car for a full week and drives 462 miles?

____

11. The sum of two numbers is 63.

a. Write a mathematical expression for the product of the two numbers as a function of one of them (x).

____

b. Use the function expression found above to find the product if one of the numbers is 7.

____

12. The sum of a number and three times another one is 96.

a. Write a mathematical expression for the product of the two numbers as a function of the first one (x).

____

b. Use the functional expression found above to find the product if the first number is 15.

____

13. A path 3 ft wide has to be built around a circular pond of radius x.

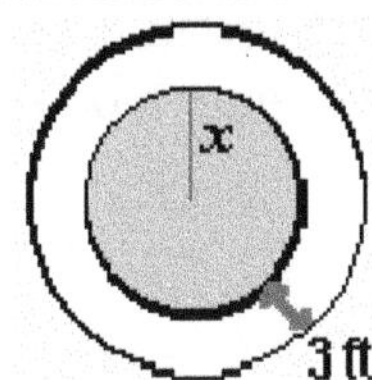

a. Find an expression for the area to be covered as a function of the pond's radius.

____

b. Use the expression obtained above to find the area that a 3ft-wide path covers around a pond with a 20ft radius.

____

14. A path 3 feet wide is to be built around the garden of the shape of a rectangle with two semicircles at the ends, as shown in the following figure.

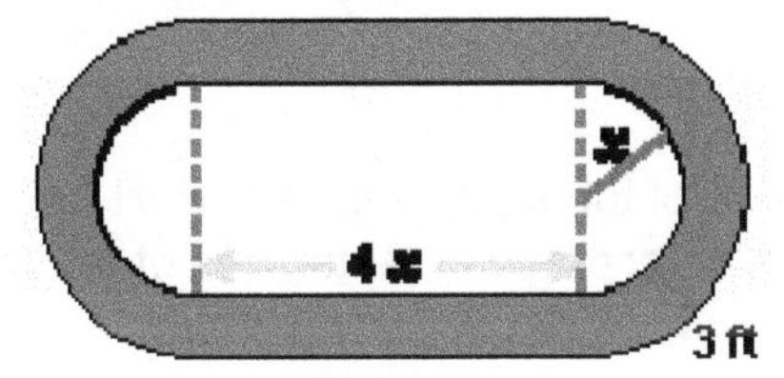

a. Find an expression for the area to be covered by the path as a function of the length x.

____

b. Use the expression found above to find the area covered by the path when the radius x is 54ft.

____

15. A person has 160 feet of chicken wire to fence a rectangular garden of width x by her house. She plans to use one of the house's wall as one of the sides of the garden, and use the chicken wire for the other three.

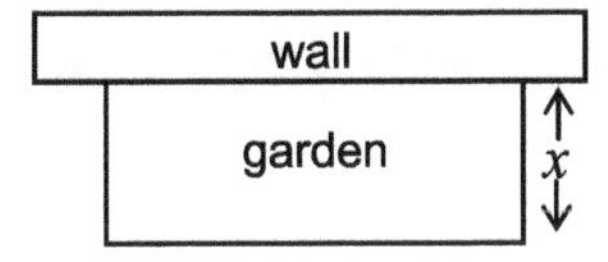

a. Find the area of the garden as a function of the garden's width x.

____

b. Use the expression found above to find the area of the garden when $x = 30$ ft.

____

16. A farmer needs to fence a $250\ \text{ft}^2$ rectangle of property for organic gardening. He is planning to fence the one side that is exposed to the wind with a taller and more resistant fence than the one he will use for the other three sides. The taller fence costs $\$15$ per foot, while the less expensive one costs $\$2.50$ per foot.

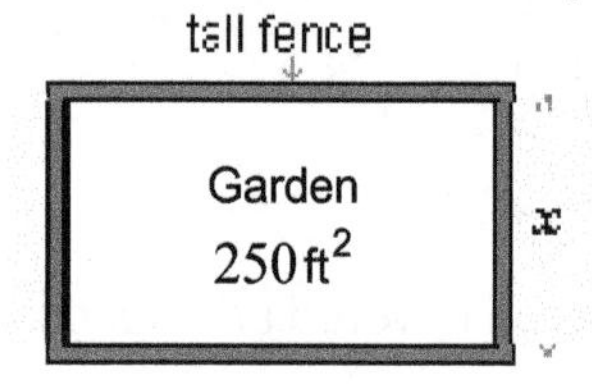

a. Write an expression for the total cost of the fence as a function of the garden's width x.

____

b. Use the expression obtained above to find the cost of fencing the organic garden if $x = 50$ ft.

____

17. A potter sells, on the average, 32 ceramic pots at the fair when he charges $\$6$ per pot. He finds that when he charges $\$7$ per pot his sales drop to 29 pots, on average. Write an expression for the potter's revenue as a function of the dollar price increase x.

____

18. A car rental agency has 24 identical cars. The owner of the agency finds that all the cars can be rented at a price of $\$25$ per day. However, for each $\$2$ dollar increase in rental, one of the cars is not rented. Find an expression for the agency's revenue as a function of the dollar increase.

____

19. An **open** box with a square base has a volume of $750\ \text{cm}^3$.

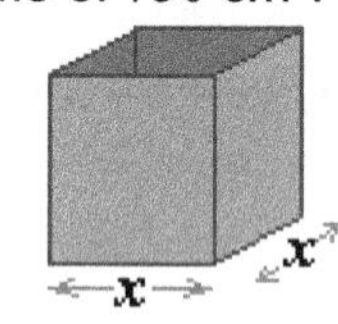

a. Find the expression of the total surface area of the box as a function of the length (x) of the side of the square base.

____

b. Use the expression found above to find the surface area of the box if $x = 15$ cm.

____

20. An **open** cylindrical container of radius x and height equal to 3 times its diameter will be made from pieces of cardboard.

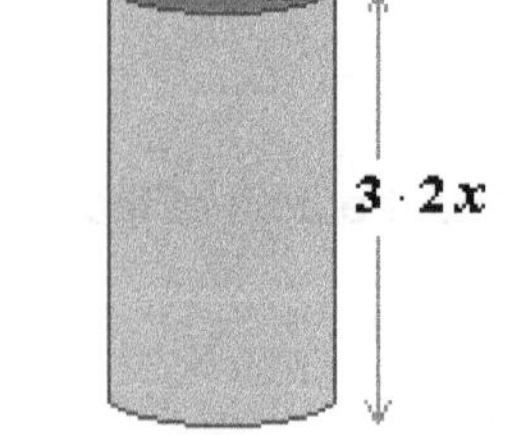

a. Find an expression for the total surface area of cardboard to be used as a function of the cylinder's radius.

____

continued...

b. Use the expression found above to find the surface area of the cylinder if $x = 12.5$ in.

____

21. A piece of wire 6 feet long is to be bent into the shape of a rectangle. Find an expression for the area of the rectangle made with the wire as a function of the length x of one of the rectangle's sides.

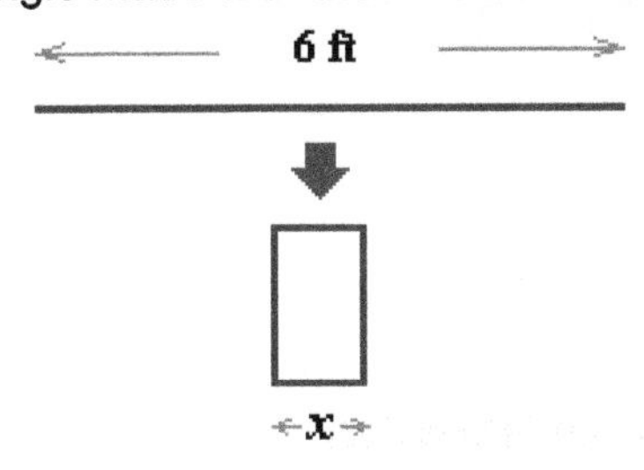

____

22. A piece of wire 15 inches long is to be cut in two parts, and each piece to be bent into the shape of a circle. Find an expression for the total surface area of the two circles made with the pieces of wire as a function of the length x of one of the pieces.

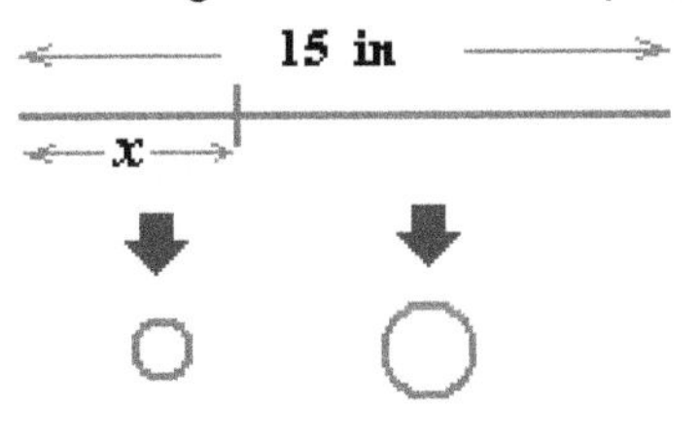

____

Chapter 1 Section 5: Operations with Functions

Practice Problems

Pairs of the graph of a function and a transformation of it are given below. For each pair, indicate what transformation was performed to produce the second graph, and express it in function notation.

Original Function | **Transformed Function**

1.

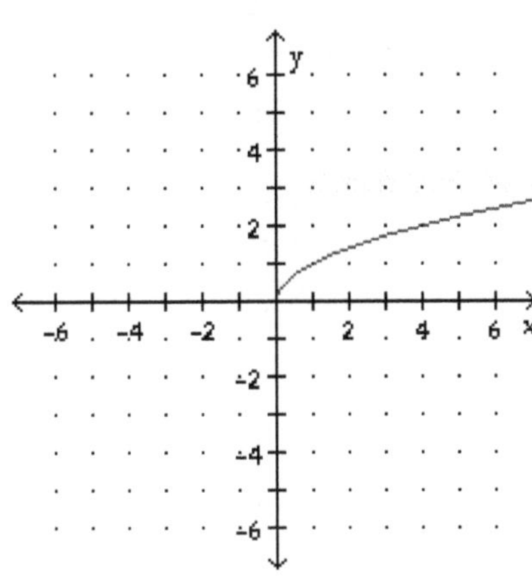

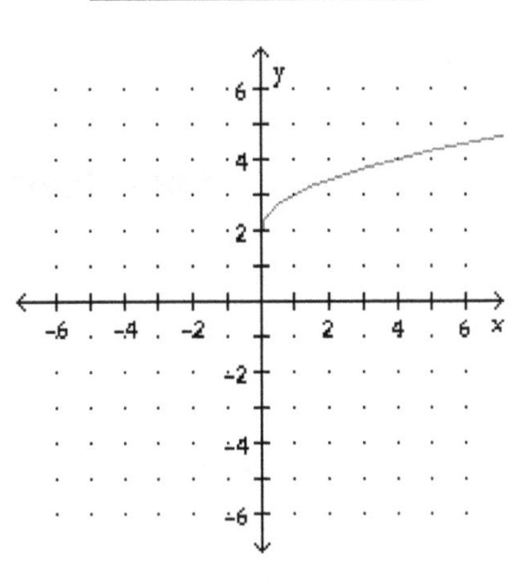

For example:
The constant 2 was added to the function producing a vertical shift: $f(x) \Rightarrow f(x)+2$.

2.

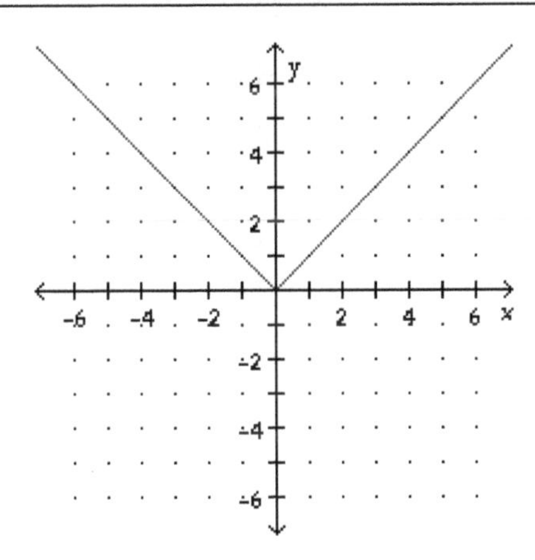

3.

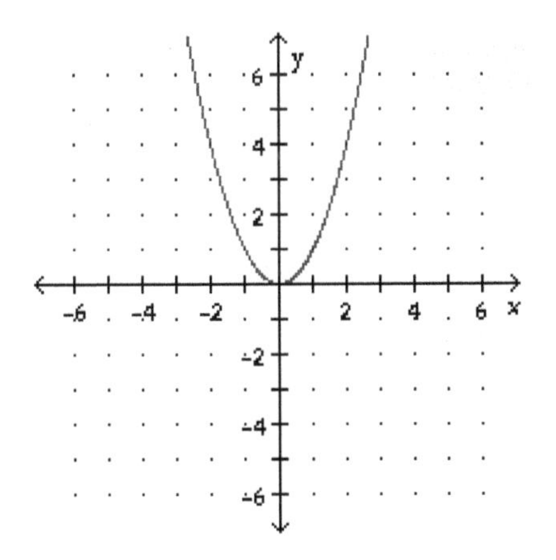

	Original Function	**Transformed Function**

4.

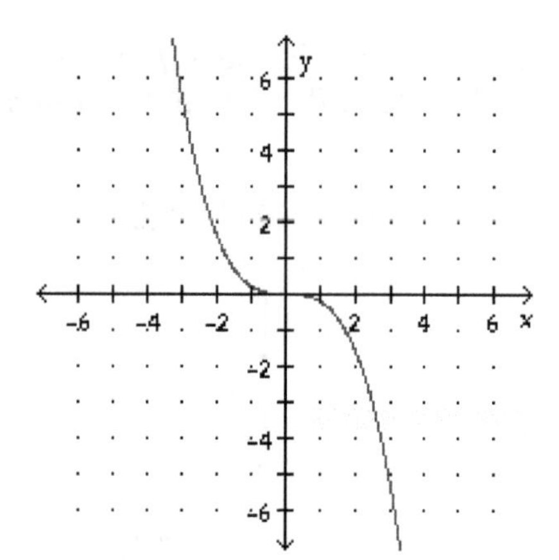

5.

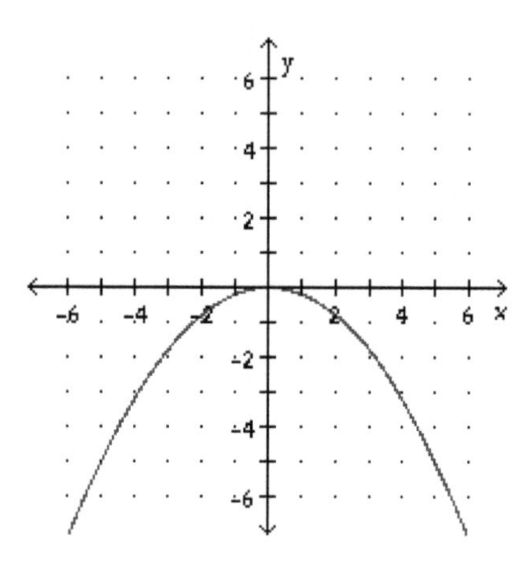

6.

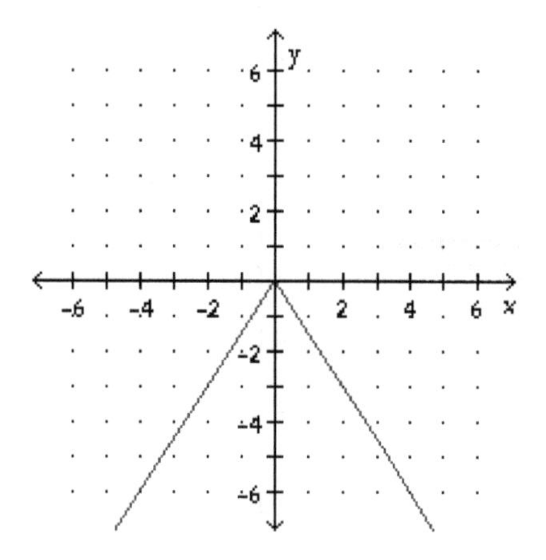

Original Function | **Transformed Function**

7.

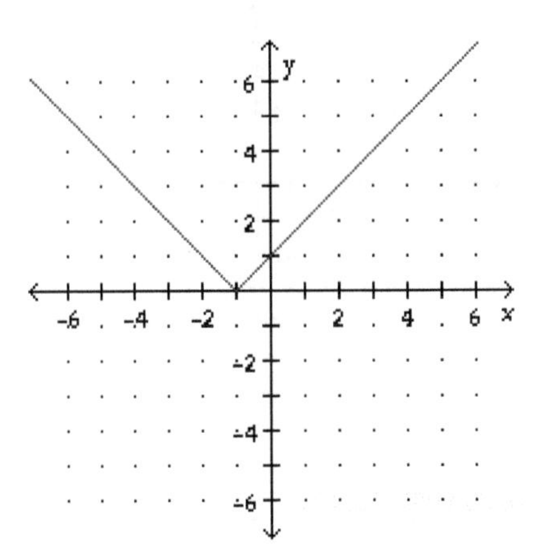

8.

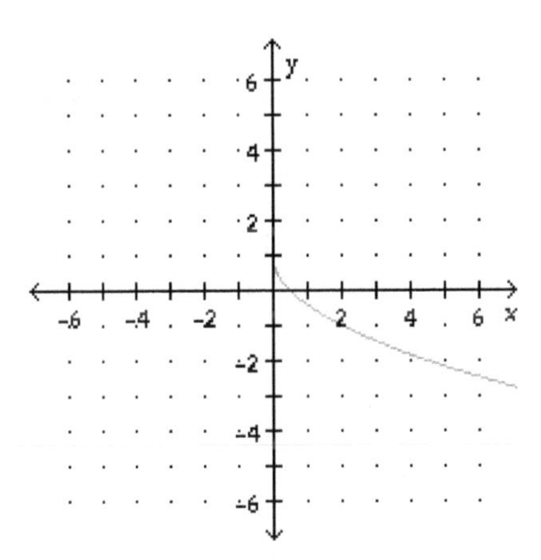

9.

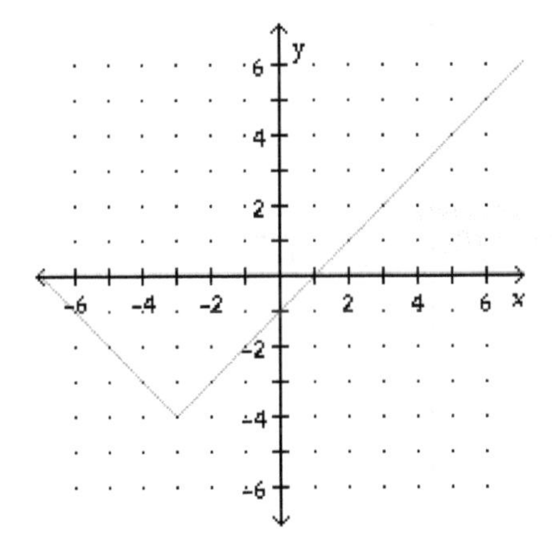

10.

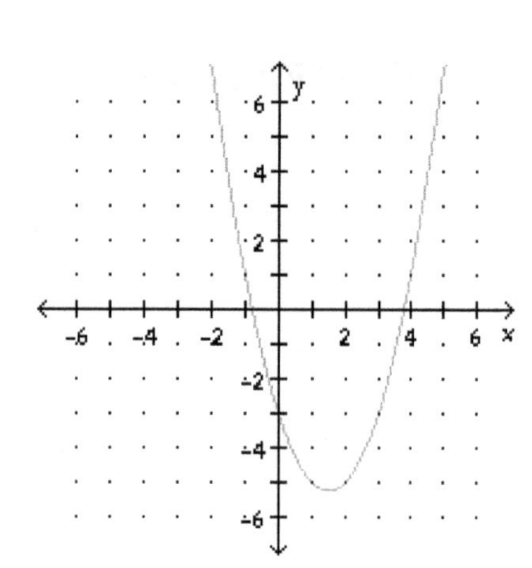

11.

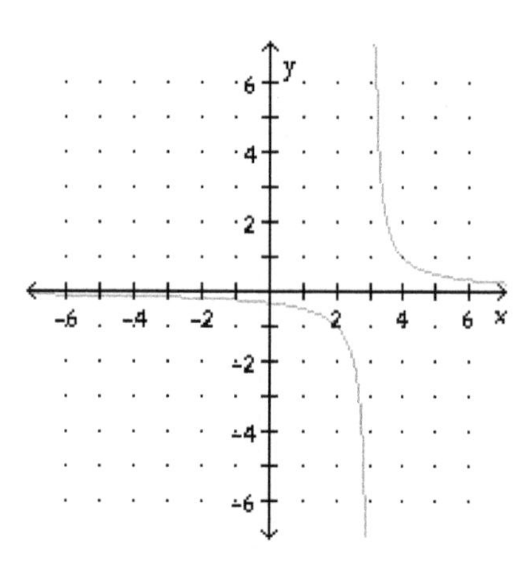

12. *The graph of a function, $y = f(x)$, is given below. Use what you learned about the transformation of functions to create the graph of $y = -f(x+2)$.*

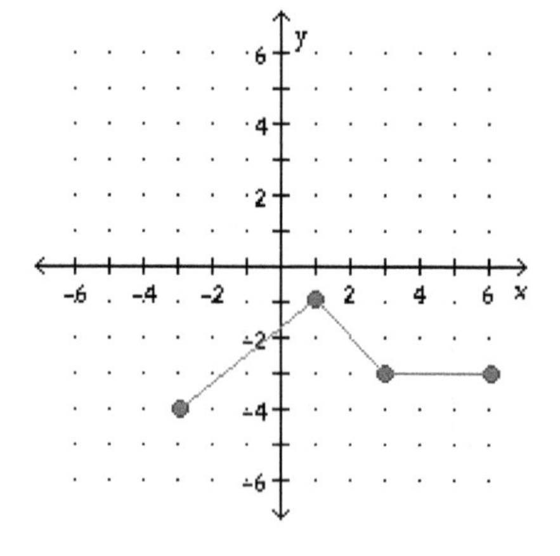

13. *The graph of a function, $y = f(x)$, is given below. Use what you learned about the transformation of functions to make the graph of $y = f(-x) - 3$.*

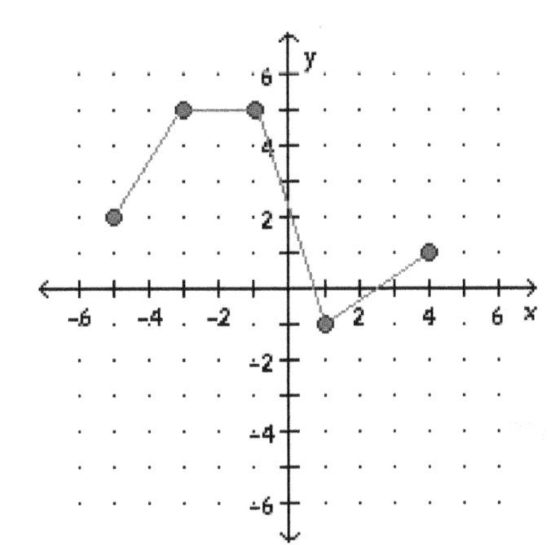

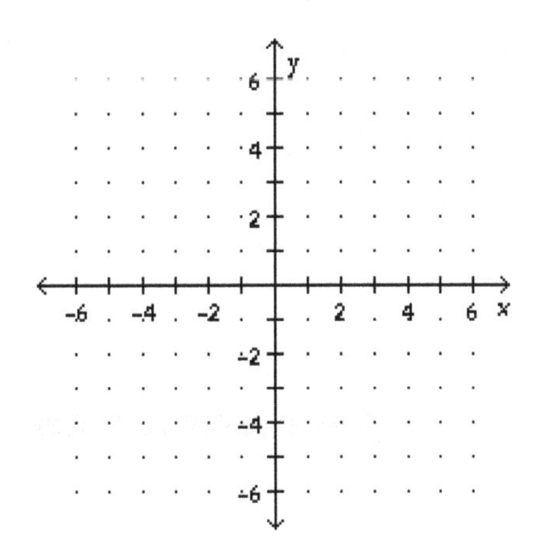

14. *Use what you learned about the transformation of functions to write the algebraic expression of a function whose graph has the shape of each of the following:*

a. $y = x^2$, but shifted three units to the right and one unit down.

____

b. $y = \sqrt{x}$, but shifted to the left 4 units and stretched vertically by a factor of 2.

____

c. $y = (x+7)^3 - 4$, but shifted to the right 2 units, down 3 units, and then reflected about the x-axis.

____

d. $y = \dfrac{x-4}{x^2}$, but shifted to the left 2 units and shrunk vertically by a factor of three.

____

e. $y = \sqrt[3]{x^2 - x + 5}$, but shifted to the right one unit, then reflected about the y-axis, and then shrunk vertically by a factor of 2.

____

Use the function properties for even functions [$f(-x) = f(x)$] and for odd functions [$f(-x) = -f(x)$], to identify which of the following are odd and which are even.

15. $f(x) = -\frac{1}{x^6}$ ____________

16. $f(x) = \frac{4x^3}{5}$ ____________

17. $f(x) = \frac{1}{2}x - \frac{1}{2}$ ____________

18. $f(x) = 3x$ ____________

19. $f(x) = \frac{x^2}{5}$ ____________

20. $f(x) = \frac{1}{2x}$ ____________

21. $f(x) = x^3 + 4x$ ____________

22. $f(x) = x^3 - 3x + 1$ ____________

23. $f(x) = 3x^4 - 5x^2$ ____________

24. $f(x) = 2x^3 + 4$ ____________

25. $f(x) = \frac{1}{x - 4}$ ____________

26. $f(x) = \frac{3}{x^4 - 5x^2}$ ____________

Find the domain of each function $h(x)$ by studying the domains of $f(x)$ and $g(x)$.

27. $h(x) = f(x) + g(x)$ where $f(x) = \sqrt{x-3}$ and $g(x) = x^2 - 4$

____

28. $h(x) = f(x) + g(x)$ where $f(x) = \dfrac{1}{\sqrt{5-x}}$ and $g(x) = \dfrac{1}{2-x}$

____

29. $h(x) = f(x) - g(x)$ where $f(x) = \dfrac{2}{2x-1}$ and $g(x) = \dfrac{x}{x+4}$

____

30. $h(x) = f(x) \cdot g(x)$ where $f(x) = \dfrac{2}{x-2}$ and $g(x) = \dfrac{x}{x+2}$

____

31. $h(x) = f(x) \cdot g(x)$ where $f(x) = x + 3$ and $g(x) = \sqrt{2x+5}$

____

32. $h(x) = \dfrac{f(x)}{g(x)}$ where $f(x) = \sqrt{4-3x}$ and $g(x) = \dfrac{x}{x-1}$

____

Find the inverse of each of the following functions.

33. $f(x) = 3x - 7$ ____________________

34. $f(x) = \dfrac{-5x+2}{3}$ ____________________

35. $f(x) = 2x^3 + 1$ ____________________

36. $f(x) = \sqrt[3]{3x-2}$ ____________________

37. $f(x) = \dfrac{3}{5-2x} - 4$ ____________________

38. $f(x) = \dfrac{x+5}{2-x}$ ____________________

39. $f(x) = \dfrac{2-x}{3x-4}$ ____________________

Find the composition of the functions indicated below.

40. $f \circ g(x)$ where $f(x) = x - 2$ and $g(x) = 2x^2 + 5$ ________________

41. $g \circ f(x)$ where $f(x) = x - 2$ and $g(x) = 2x^2 + 5$ ________________

42. $g \circ g(x)$ where $g(x) = 2x^2 + 5$ ________________

43. $f \circ g(x)$ where $f(x) = \dfrac{1}{x+2}$ and $g(x) = x^2 - 7x + 2$ ________________

44. $f \circ g(x)$ where $f(x) = \sqrt[3]{x-7}$ and $g(x) = 1 - x^3$ ________________

45. $g \circ f(x)$ where $f(x) = x + 5$ and $g(x) = \dfrac{x-2}{x+3}$ ________________

46. $g \circ f(x)$ where $f(x) = 2x - 1$ and $g(x) = \dfrac{x^2 - 1}{x+2}$ ________________

Tables of input-output values for two functions, $f(x)$ and $g(x)$, are given below. Although the functional expressions are not given, use your knowledge of operations with functions to find the requested values.

x	$f(x)$
−1	5
0	4
1	3
2	0
3	2
4	−1
5	1

x	$g(x)$
−1	2
0	3
1	4
2	−1
3	1
4	5
5	0

47. $3 \cdot f(2)$ ________________

48. $-5 \cdot g(0)$ ________________

49. $(f+g)(3)$ ________________

50. $(f-g)(2)$ ________________

51. $(g-f)(1)$ ________________

52. $(f \cdot g)(-1)$ ________________

53. $\left(\dfrac{f}{g}\right)(4)$ ________________

54. $f \circ g(1)$ ________________

55. $g \circ f(5)$ ________________

56. $g \circ g(4)$ ________________

57. $f \circ f(2)$ ________________

58. $f^{-1}(3)$ ________________

59. $g^{-1}(0)$ ________________

60. $f\left(g^{-1}(3)\right)$ ________________

A list of functions and their inverses are given below. Use the property of the composition of inverse functions: $f^{-1}(f(x)) = f\left(f^{-1}(x)\right) = x$ *to verify that they are the inverse of each other.*

61. $f(x) = \dfrac{1}{2x}$ and $f^{-1}(x) = \dfrac{1}{2x}$ ________

62. $f(x) = \dfrac{x+1}{2}$ and $f^{-1}(x) = 2x - 1$ ________

63. $f(x) = \dfrac{x-3}{2}$ and $f^{-1}(x) = 2x + 3$ ________

64. $f(x) = \dfrac{4x-3}{2-x}$ and $f^{-1}(x) = \dfrac{2x+3}{x+4}$ ________

65. $f(x) = x^3 + 5$ and $f^{-1}(x) = \sqrt[3]{x-5}$ ________

66. $f(x) = \dfrac{3-x}{x+1}$ and $f^{-1}(x) = \dfrac{3-x}{x+1}$ ________

67. $f(x) = 2x^3 + 1$ and $f^{-1}(x) = \sqrt[3]{\dfrac{x-1}{2}}$ ________

68. $f(x) = \dfrac{2-x}{3x-4}$ and $f^{-1}(x) = \dfrac{2+4x}{3x+1}$ ________

Find each requested composition of the given functions:

$f(x) = \sqrt{x^2+1}$ *and* $g(x) = \dfrac{3x^2-5}{2}$.

69. $f \circ g(1)$ __________

70. $f \circ g(-3)$ __________

71. $f \circ f(0)$ __________

72. $g \circ f(2)$ __________

73. $g \circ g(3)$ __________

CONTENT
on Demand

Precalculus 2nd Edition

Practice Problem Worksheets

Chapter 2: Linear & Quadratic Functions

Chapter 2 Section 1: An Overview of Linear Functions

Practice Problems

1. Find the constant rate relationship between the variables in each example given below. Graph each direct proportion as a line passing through $(0,0)$.

a. A farmer uses 30 pounds of mole repellant to treat 5 acres of land.

____

b. A recipe calls for 10 oz of butter for every 4 cups of flour.

____

c. At a party, 30 guests consumed 46 cans of soda.

____

d. Six chicken lay an average of 28 eggs per week.

____

e. A person needs 5 cans of adhesive to glue 270 ft^2 of linoleum.

____

2. The cost of producing 1000 cans of soup is $890, while the cost of producing 5500 cans of soup is $4290. Find the rate of change of the production cost per can of soup produced.

____

3. The cost of producing 1000 CD's is $540, while the cost of producing 5000 CD's is $2100. Find the rate of change of the production cost per CD produced.

____

4. The cost of cleaning a 45-office building is $42,300 per month, while the cost of cleaning a 124-office building is $96,720. Find the rate of change of the cleaning cost per month per individual office.

____

5. The cost of producing 58 handcrafted bracelets is $126, while the cost of producing 125 bracelets is $222. Find the rate of change of the production cost per bracelet produced.

____

6. The cost of upgrading 346 computers is $17,992, and the cost of upgrading 610 computers is $31,720. Find the rate of change of the upgrading cost per computer.

____

7. At the beginning of the year, the balance of a person's bank account was $1435. Six months later the balance in the account was $2365. Find the rate of change of the account's balance per month.

____

8. At the beginning of the year, a person's savings was $4350. Eleven months later his savings are $6430. Find the rate of change of the person's savings per month.

____

9. A farm had 6 rabbits at the beginning of the year. Nine months later, the number of rabbits increased to 18. Find the rate of change in the rabbit population per month.

____

10. A coin collector started with 23 coins, and 7 years later his collection has increased to 178 pieces. Find the rate of change in number of coins per year.

____

11. There were 280 owls in a national forest. In 8 years, their number has fallen to 126. Find the rate of change in number of owls per year.

____

12. Find the slope of the linear function through each of the following pairs of points on the x-y plane:

a. $(3,2)$ and $(0,7)$

____

b. $(-1,5)$ and $(3,5)$

____

c. $(-2,2)$ and $(-2,-4)$

---

d. $\left(1,\frac{2}{3}\right)$ and $\left(-4,\frac{7}{3}\right)$

---

e. $\left(\frac{1}{4},\frac{2}{5}\right)$ and $\left(-\frac{1}{2},-1\right)$

---

13. Find an equation of the line which goes through each pair of points on the x-y plane:

a. $(2,1)$ and $(-2,5)$

---

b. $(-1,3)$ and $(-3,-2)$

$f(x)=\frac{5}{2}x+\frac{11}{2}$

---

c. $(-4,3)$ and $\left(\frac{3}{2},3\right)$

---

d. $\left(-\frac{1}{3},2\right)$ and $\left(-\frac{1}{3},\frac{2}{5}\right)$

---

e. $\left(-\frac{3}{5},-3\right)$ and $\left(\frac{2}{5},-2\right)$

---

14. The following tables show the x and y coordinates of points that belong to certain linear functions $f(x)$. Find the missing coordinate values.

a.

x	$f(x)$
0	3
1	?
2	9

b.

x	$f(x)$
0	−3
−3	?
6	−9

c.

x	$f(x)$
0	−3
?	0
5	−1

d.

x	$f(x)$
?	−1
−6	0
4	$-\frac{5}{3}$

e.

x	$f(x)$
?	1
10	0
2	$\frac{8}{5}$

15. A pizza place charges $56 to rent a "Birthday Party" room, and an additional $5.50 per child attending the party. Find the total charged for a birthday party of 36 children.

____

16. A plumber charges $55 per house call plus $68 per hour of labor. Find the total charges for a house call where the plumber worked for 2.5 hours.

____

17. A stucco company charges a fixed rate of $458 as a set-up fee, plus $72 per hour of labor. Find the total cost of a job that took 71 hours to complete.

____

18. A phone company charges a fixed rate of $5.95 per month and 15 cents per minute used. Find the total billed to a person under this plan who used 134 minutes in a month.

____

19. A community college charges a campus fee of $75 per semester, plus $28 per unit taken by a student. Find the total cost for a student taking 15 units the first semester and 12 units the second semester.

____

20. A new washing machine costs $560. In four years its value will have decreased to $250. Assuming that the decrease is linear, find the value of the washing machine three years after its purchase.

____

21. The price of a computer chip has decreased from $4.50 to $2.70 in 3 years. Assuming its value decreases linearly, find the price of the chip next year.

____

22. The balance in a person's bank account has decreased from $870 to $525 in 5 months. If this linear decrease continues, find what the balance would be in another 2 months.

____

23. A new car cost $23,560. In four years, its value has decreased linearly to $16,580. If its value keeps on following this linear decrease, find the car's value in three more years.

____

24. A box of 12 light bulbs was bought by a homeowner as spare supply. If 2 bulbs are left in the box 5 years later, and assuming that the usage was linear, how many spare bulbs where there in the box 2 years ago?

____

Chapter 2 Section 2: Geometry of Linear Functions

Practice Problems

Find the y-intercept and the zero crossing of each of the following linear functions.

1. $f(x)=3x+2$

y-intercept: ________

zero crossing: ________

2. $f(x)=-x-5$

y-intercept: ________

zero crossing: ________

3. $f(x)=-2x+6$

y-intercept: ________

zero crossing: ________

4. $f(x)=4x-7$

y-intercept: ________

zero crossing: ________

5. $f(x)=\frac{1}{2}x+3$

y-intercept: ________

zero crossing: ________

6. $f(x)=2x+\frac{3}{4}$

y-intercept: ________

zero crossing: ________

7. $f(x)=-\frac{2}{3}x-\frac{5}{3}$

y-intercept: ________

zero crossing: ________

8. $f(x)=\frac{4}{5}x+\frac{1}{5}$

y-intercept: ________

zero crossing: ________

9. $f(x)=-\frac{2}{5}x-\frac{5}{3}$

y-intercept: ________

zero crossing: ________

10. $f(x)=\frac{5}{4}x+\frac{3}{8}$

y-intercept: ________

zero crossing: ________

11. $f(x)=-3$

y-intercept: ________

zero crossing: ________

Find the equation of the line parallel to each given linear function and going through the indicated point. Give your answer in slope-intercept form, $f(x)=mx+b$, and graph both lines and the given point.

12. Parallel to $f(x)=2x-4$ and through point $(1,4)$.

13. Parallel to $f(x) = x + 5$ and through the point $(-1, 2)$.

14. Parallel to $f(x) = -x - 8$ and through point $(3, -4)$.

15. Parallel to $f(x) = -\frac{1}{2}x - 3$ and through point $(2, 5)$.

16. Parallel to $f(x) = \frac{5}{3}x + \frac{1}{3}$ and through point $(-6, 1)$.

17. Parallel to $f(x)=\frac{1}{4}x-\frac{7}{4}$ and through point $(-3,1)$.

18. Parallel to $f(x)=-\frac{3}{2}x-4$ and through point $\left(4,\frac{3}{2}\right)$.

19. Parallel to $f(x)=\frac{3}{5}x+\frac{11}{2}$ and through point $\left(-2,-\frac{7}{5}\right)$.

20. Parallel to $f(x) = -\frac{4}{3}x + \frac{12}{5}$ and through point $\left(\frac{1}{2}, \frac{7}{6}\right)$.

21. Parallel to $f(x) = -\frac{3}{2}x - \frac{11}{4}$ and through point $\left(-\frac{1}{3}, \frac{7}{2}\right)$.

22. Parallel to $f(x) = 7$ and through point $(3, 2)$.

23. Parallel to $x = -4$ and through point $(2,-1)$.

Find the equation of the line perpendicular to each given linear function and going through the indicated point. Give your answer in slope-intercept form, $f(x) = mx + b$, and graph both lines and the given point.

24. Perpendicular to $f(x) = 2x - 4$ and through point $(1,4)$.

25. Perpendicular to $f(x) = x + 5$ and through the point $(-1,2)$.

26. Perpendicular to $f(x) = -x - 8$ and through point $(3, -4)$.

27. Perpendicular to $f(x) = -\frac{1}{2}x - 3$ and through point $(2, 5)$.

28. Perpendicular to $f(x) = \frac{5}{3}x + \frac{1}{3}$ and through point $(-5, 1)$.

29. Perpendicular to $f(x) = \frac{1}{4}x - \frac{7}{4}$ and through point $(-2, 1)$.

30. Perpendicular to $f(x) = -\frac{3}{2}x - 4$ and through point $\left(3, \frac{3}{2}\right)$.

31. Perpendicular to $f(x) = \frac{3}{5}x + \frac{11}{2}$ and through point $\left(-3, -\frac{7}{3}\right)$.

32. Perpendicular to $f(x) = -\frac{4}{3}x + \frac{12}{5}$ and through point $\left(\frac{2}{3}, \frac{7}{8}\right)$.

33. Perpendicular to $f(x) = -\frac{3}{2}x - \frac{11}{4}$ and through point $\left(-\frac{3}{4}, \frac{7}{6}\right)$.

34. Perpendicular to $f(x) = 7$ and through point $(3,2)$.

35. Perpendicular to the vertical line $x = -4$ and through point $(2,-1)$.

Find the point of intersection of the following lines. Express your answer as an (x, y) pair, and graph the lines and the intersection point.

36. $f(x) = -2x + 5$ and $g(x) = 3x - 5$

37. $f(x) = 3x + 13$ and $g(x) = x + 7$

38. $f(x) = -2x$ and $g(x) = \frac{2}{3}x - \frac{8}{3}$

39. $f(x) = x - 1$ and $g(x) = \frac{3}{2}x$

40. $f(x) = 2x + 3$ and $g(x) = \frac{5}{3}x + \frac{10}{3}$

41. $f(x) = \frac{1}{2}x + 3$ and $g(x) = \frac{5}{3}x + 10$

42. $f(x) = 3x + \frac{11}{2}$ and $g(x) = -\frac{4}{3}x + \frac{10}{3}$

43. $f(x) = -\frac{7}{6}x + \frac{1}{2}$ and $g(x) = \frac{4}{3}x + 3$

Practice Problems

Use the linear regression tool in Calculator on Demand to find the linear function that best fits the data shown in each of the following tables. Give the linear function in slope y-intercept form, the correlation coefficient, and the coefficient of determination.

1.

x	0	1	2	3	4
y	2	5	6	8	10

____

2.

x	0	2	4	6	8
y	–1.4	2.1	5.4	7.1	9.3

____

3.

x	–4	–2	0	2	4
y	10.2	22.1	31.4	40.6	52.0

____

4.

x	1.2	3.5	6.3	8.9	10.4
y	5.3	3.7	2.1	1.4	0.3

____

5.

x	0.1	2.3	5.7	6.8	9.4
y	120	108	72	64	31

____

6. The following table shows the world's energy consumption of hydroelectric power (in quadrillions of BTU's) for the given years. Use linear regression to find the best linear function that models the data. Use 1950 as year number "0," so 1954 will correspond to year #4, 1958 to year #8, etc.

year	1954	1958	1962	1966	1970	1974
BTU's	1.39	1.63	1.82	2.07	2.65	3.31

____

7. The following table shows the world's energy consumption of petroleum (in quadrillions of BTU's) for the given years. Use linear regression to find the best linear function that models the data. Use 1950 as year number "0," so 1954 will correspond to year #4, 1958 to year #8, etc.

year	1954	1958	1962	1966	1970	1974
BTU's	15.84	18.53	21,05	24.40	29.52	33.45

____

8. The following table shows the duration of around-the-world trips from New York to New York in different years as the means of transportation and average cruise speeds improved with time. Use linear regression to find the best linear function that models the data. Use the year 1880 as the year "zero."

year	*duration*
1889	72 *days*
1890	67 *days*
1926	28 *days*
1931	8 *days*
1933	115 *hours*
1938	3 *days* 19 *hours*
1963	46.5 *hours*

____

9. The following table shows OPEC's production of crude oil (in millions of barrels per day) for the given years. Use 1960 as year number "0." Answer the following questions.

year	1960	1964	1968	1972	1976
barrels	8.7	12.98	18.97	27.09	30.74

a. Use linear regression to find the linear function that best models the data.

____

b. What would have been the prediction of this model for the year 1990?

____

c. The actual OPEC production for the year was 1990 was 23.46 million barrels per day, much lower that the prediction for a linear function based on the given data. Find the new linear function that models the data if you add this information for the year 1990 to the list.

____

Chapter 2 Section 3: An Overview of Quadratic Functions

Practice Problems

Find the mathematical expression for the transformation of each of the following quadratic functions that moves the vertex to a specified new location on the x-y plane. Express your answer in vertex form, and graph the original parabola and its transformation (on the same plane).

1. $f(x) = x^2$ to the location $(3,7)$

2. $f(x) = -x^2$ to the location $(-5,0)$

3. $f(x) = -2x^2$ to the location $(1,-4)$

4. $f(x) = 3x^2 + 1$ to the location $(-3,2)$

5. $f(x) = -4(x-2)^2 + 5$ to the location $(-2,2)$

Find the mathematical expression for the transformation of each of the following quadratic functions that moves the vertex according to a given sequence of shifts. Express your answer in vertex form, and graph the original parabola and its transformation (on the same plane).

6. $f(x) = 3x^2$, 2 units to the right and 5 units down.

7. $f(x) = 2x^2 - 6$, 3 units to the left and 4 units up.

8. $f(x) = -(x-1)^2 + 2$, 4 units to the left and 5 units down.

9. $f(x) = 5(x+2)^2 + 4$, 5 units to the right and 6 units down.

10. $f(x) = -3(x-5)^2 - 6$, 4 units to the left and 4 units up.

Expand each quadratic function given in vertex form into standard form.

11. $f(x) = (x-3)^2 + 1$ ________________

12. $f(x) = -(x+2)^2 + 3$ ________________

13. $f(x) = 2(x-5)^2 - 7$ ________________

14. $f(x) = -2(x+3)^2 + 8$ ________________

15. $f(x) = -3(x-4)^2 + 12$ ________________

Use the formulas given in the Lesson to convert each of the following quadratic functions from standard form into vertex form.

16. $f(x) = x^2 + 4x + 3$ ________________

17. $f(x) = -x^2 - 2x + 4$ ________________

18. $f(x) = 2x^2 + 12x + 10$ ________________

19. $f(x) = -3x^2 - 24x - 31$ ________________

20. $f(x) = -2x^2 + 20x - 18$ ________________

Locate the minimum value produced by each function given below whose graph is that of a concave-up parabola, and give its function expression in vertex form.

21. $f(x) = x^2 + 2x - 3$ ________________

22. $f(x) = x^2 - 4x + 5$ ________________

23. $f(x) = 2x^2 - 8x - 7$ ________________

24. $f(x) = 6x^2 + 6x + 5$ ________________

25. $f(x) = 3x^2 + 5x - 2$ ________________

Locate the maximum value produced by each function given below whose graph is that of a concave-down parabola, and then give its function expression in vertex form.

26. $f(x) = -x^2 - 4x + 3$

____

27. $f(x) = -x^2 + 5x - 6$

____

28. $f(x) = -2x^2 + x + 5$

____

29. $f(x) = -3x^2 - 9x + 5$

____

30. $f(x) = -4x^2 + 10x - 3$

____

31. A wholesale merchant sells 20 cases of bottles of olive oil per day when he charges $110 per case. His sales decrease by one case per day when he increases the price to $115. Find the price he should charge per case in order to maximize sales, and also calculate the maximum sales amount.

____

32. A person sells 140 herbal soaps per day at the farmer's market when she prices each soap at $1.20. When she charges $1.40 per soap, her per-day sales decrease to 125. Find the price she should be charging per soap in order to maximize sales, and also calculate the maximum sales amount.

____

33. A book reseller sells 150 books per week when he charges $18 per book. When he lowers the price of each book to $16 he sells 175 books per week. Find the price he should charge for the books in order to maximize sales, and also calculate the maximum sales amount.

____

34. The manager of a 75-apartment complex rents 50 apartments when he charges rent of $1200 per month. When he raises the rent to $1300, he rents only 44 apartments. Find the rental fee per apartment he should charge per month in order to maximize his income, and also calculate the maximum income. How many apartments would be rented under this condition?

____

35. A person has 160 feet of chicken wire to fence a rectangular garden of width x next to her house. She plans to use one of the house's walls as one of the sides of the garden, and use the chicken wire for the other three sides. Find the size of the garden that will give the maximum area. What is that area?

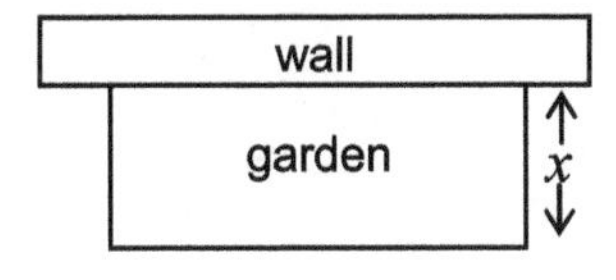

____

36. A piece of wire 6 feet long is to be bent into the shape of a rectangle.

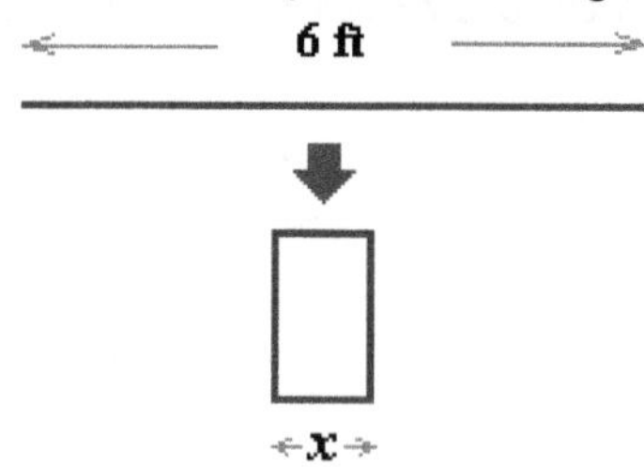

Find the dimensions of the rectangle that can be made with the piece of wire that gives the maximum possible area. What is that area?

____

37. A baseball is thrown through the air. Its height with respect to the ground (in feet) is described by the expression: $f(t) = -16t^2 + 62t + 5.5$ with t describing the time in seconds. Find the maximum height that the ball reaches.

____

38. The addition of two numbers is 18. Find the maximum possible product of the two.

____

39. The addition of a number and 4 times another number is 104. Find the maximum possible product of the two.

____

40. The population of a certain rain forest species has followed the following quadratic expression for 40 years: $Population(x) = -0.002x^2 + 0.08x + 850$, where x is the time in years, counting from 1960. When did the population reach its maximum value? What was that maximum value?

____

41. A manufacturing company finds that the cost in dollars (C) of producing x units of one of their products, is given by the function: $C(x) = 3x^2 - 1{,}200x + 124{,}350$. Find the number of units to be produced to minimize the cost, and find the minimum cost.

____

Chapter 2 Section 4: Geometry of Quadratic Functions

Practice Problems

Find the equation of the axis of symmetry of the parabolas described by the following quadratic functions.

1. $f(x) = x^2 - 3x$ ____________

2. $f(x) = x^2 + 4x - 2$ ____________

3. $f(x) = 2x^2 - 9x$ ____________

4. $f(x) = -3x^2 - 5x + 2$ ____________

5. $f(x) = -4x^2 + x - 3$ ____________

Find the x-intercepts (if they exist) of each of the following quadratic functions. Graph the function and mark the intercepts.

6. $f(x) = x^2 - 36$

7. $f(x) = x^2 - 6$

8. $f(x) = -x^2 - 2$

9. $f(x) = -2x^2 + 5x$

10. $f(x) = 3x^2 + 4x$

11. $f(x) = x^2 + 3x - 4$

12. $f(x) = x^2 + 5x - 4$

13. $f(x) = x^2 - 4x + 4$

14. $f(x) = 2x^2 - 5x - 12$

15. $f(x) = 2x^2 + 5x - 10$

16. $f(x) = 4x^2 - 12x + 10$

17. $f(x) = -2x^2 - 3x + 10$

Analyze the discriminant associated with the following quadratic equations to decide the type and number of solutions you expect. Do not solve the equations.

18. $x^2 - 6x + 9 = 0$ ____________________

19. $x^2 - x - 6$ ____________________

20. $2x^2 - 3x + 7 = 0$ ____________________

21. $2x^2 + 5x + 1 = 0$ ____________________

22. $4x^2 - 13x - 12 = 0$ ____________________

The following problems involve projectile motion. The motion of objects under the effect of gravity when air resistance is neglected is well described by parabolas. We often find the gravitational constant in the expression of a trajectory. The quantity 32 (when the gravitational constant is given in feet/second ^{2}) or 9.8 (when given in meters/second ^{2}) may appear explicitly in the trajectory expression.

23. A person throws a ball upwards with an initial velocity. The height of the ball (in meters) with respect to his hand is given by the expression: $h(t) = 15t - \dfrac{9.8\,t^2}{2}$, where t is the flying time of the ball (in seconds).

Find the following:

a. The maximum height the ball reaches.

____

b. The time it takes the ball to come back to the thrower's hand.

____

24. A person throws a rock upwards with an initial velocity. The height of the rock (in feet) with respect to his hand is given by the expression: $h(t) = 210t - \dfrac{32\,t^2}{2}$, where t is the flying time of the rock (in seconds).

Find the following:

a. The maximum height the rock reaches.

____

b. The time it takes the rock to come back to the thrower's hand.

____

25. A projectile is fired horizontally from a cliff 150 meters high. If the expression of the projectile's height with respect to the ground is given by: $h(t) = 150 - \dfrac{9.8\,t^2}{2}$, where the time t is given in seconds, how long does it take to hit the ground?

____

26. A projectile is fired horizontally from a cliff 1240 feet high. If the expression of the projectile's height with respect to the ground is given by: $h(t) = 1240 - \dfrac{32\,t^2}{2}$, where the time t is given in seconds, how long does it take to hit the ground?

____

27. A relief package is dropped from a plane. The trajectory that the package follows can be represented by: $h(x) = -0.241x^2 + 4.6$ where $h(x)$ and x are the vertical position and horizontal displacement, respectively, both given in kilometers (thousands of meters). Find the following:

a. The height the package was released from.

____

b. How far from the point where the package was released does it hit the ground?

____

c. The formula for the trajectory is a parabola; what are the x and y coordinates of its vertex and the equation of its axis of symmetry?

____

28. A relief package is dropped from a plane. The trajectory that the package follows can be represented by: $h(x) = -0.836x^2 + 16.4$ where $h(x)$ and x are the vertical position and horizontal displacement respectively, both given in thousands of feet. Find the following:

a. The height the package was released from.

____

b. How far from the point where the package was released does it hit the ground?

____

c. The formula for the trajectory is a parabola; what are the x and y coordinates of its vertex and the equation of its axis of symmetry.

____

29. A cannon ball is fired from a cliff 350 feet above the water. The height h of the projectile above the water is given by the expression $h(x) = -\frac{32}{120000}x^2 + x + 350$ where x is the horizontal distance traveled by the cannon ball from the base of the cliff.

a. Find the maximum height reached by the cannon ball.

____

b. Find how far from the base of the cliff the cannon ball strikes the water. (Give your answer in feet and round it to two decimals)

____

30. A cannon ball is fired from a cliff 450 feet above the water. The height h of the projectile above the water is given by the expression $h(x) = -\frac{32}{106200}x^2 + x + 450$ where x is the horizontal distance traveled by the cannon ball from the base of the cliff.

a. Find the maximum height reached by the cannon ball.

____

b. Find how far from the base of the cliff the cannon ball strikes the water. (Give your answer in feet and round it to two decimals)

____

Find the function expression of the parabola with the specified vertex and going through the indicated point. Give your answer in vertex form.

31. Vertex at (2, 5) and through (−1, 14)

____

32. Vertex at (−3,−3) and through (4, 46)

____

33. Vertex at (−5, 1) and through (−3, 9)

____

34. Vertex at $\left(\frac{3}{2}, \frac{5}{2}\right)$ and through (0, 7)

____

35. Vertex at $(0,-4)$ and through $(1, -1)$

____

36. Vertex at $(2, 0)$ and through $(5,-18)$

____

37. Vertex at $\left(\frac{1}{2},-\frac{1}{4}\right)$ and through $\left(1,-\frac{1}{2}\right)$

____

38. Vertex at $\left(-\frac{2}{3},-1\right)$ and through $\left(\frac{2}{3},\frac{13}{3}\right)$

____

CONTENT on Demand

Precalculus 2nd Edition

Practice Problem Worksheets

Chapter 3: Polynomial & Rational Functions

Chapter 3 Section 1: Polynomial and Power Functions

Practice Problems

Decide if each given function is a polynomial function and state why or why not. If it is a polynomial function, write it in standard form and state its degree.

1. $f(x) = -3x + 2x^2 + 6x^{-1} - 4$

____

2. $f(x) = -23 - x - x^3 + \sqrt{2}\,x^2$

____

3. $f(x) = 5x^4 + 2x^2 + 3x^{\frac{1}{2}} - 5$

____

4. $f(x) = \dfrac{x^2 + 3x - 8}{x}$

____

5. $f(x) = 2\sqrt{x} - 3x^2 + 7x - 1$

____

6. $f(x) = -\dfrac{2x^2}{3} + 5x^3 - 8 + \dfrac{x}{5}$

____

7. $f(x) = 5x^3 - \dfrac{2}{\pi} - \pi x^4 - 4x$

____

8. $f(x) = -2x^4 - 3x^{\pi} + 3x^2 + 2$

____

9. $f(x) = -5$

____

10. $f(x)=2x^4-x^3+2x-5+\frac{1}{x}$

____

11. $f(x)=-2+5x$

____

12. $f(x)=\sqrt{2}$

____

Find each polynomial's leading term. Note that the polynomials are not in standard form.

13. $f(x)=-9x^2+4x-5x^3+2x^7-3$ ______

14. $f(x)=-x^2+5-3x^4-5x^3+x$ ______

15. $f(x)=-\frac{1}{2}x^2-6x-\frac{3}{4}x^6+8x^5+1$ ______

16. $f(x)=10-3x^2-x^5+7x^3+x$ ______

17. $f(x)=4x-20+\sqrt{3}x^4+3x^2-15x^3$ ______

What is the leading coefficient of each of the following polynomials?

18. $f(x)=-4x^2+2x-8x^3+3x^7-7$ ______

19. $f(x)=-2x^2+1-x^4-6x^3+4x$ ______

20. $f(x)=-\frac{3}{8}x^2-2x-\frac{1}{4}x^6+5x^5+4$ ______

21. $f(x)=x^6+5-2x^2-6x^5+10x^3+4x$ ______

22. $f(x)=x-3+\sqrt{5}x^4+5x^2-8x^3$ ______

23. $f(x)=\frac{2x-7+4x^4+3x^2-8x^3}{2}$ ______

Identify each of the following in this polynomial function: $f(x)=-3x^2+\frac{1}{3}x-6x^3-x^5+4$.

24. Degree of the polynomial: ______

25. Leading term: ______

26. Leading coefficient: ______

27. Linear term: ______

28. Constant term: ______

29. Quadratic term: ______

30. Cubic term: ______

Identify each of the following in this polynomial function: $f(x) = -\sqrt{5}x^3 - \frac{1}{2} + \frac{x^4}{3} - 4x^2$.

31. Degree of the polynomial: ________

32. Leading term: ________

33. Leading coefficient: ________

34. Linear term: ________

35. Constant term: ________

36. Quadratic term: ________

37. Cubic term: ________

Identify each of the following in this polynomial function: $f(x) = \frac{6x + 5x^4 - x^6 + 4x^5 - 4x^2}{3}$.

38. Degree of the polynomial: ________

39. Leading term: ________

40. Leading coefficient: ________

41. Linear term: ________

42. Constant term: ________

43. Quadratic term: ________

44. Cubic term: ________

Decide if the graph of each power function given below corresponds to an even integer power or to an odd integer power. Explain your answer.

45.

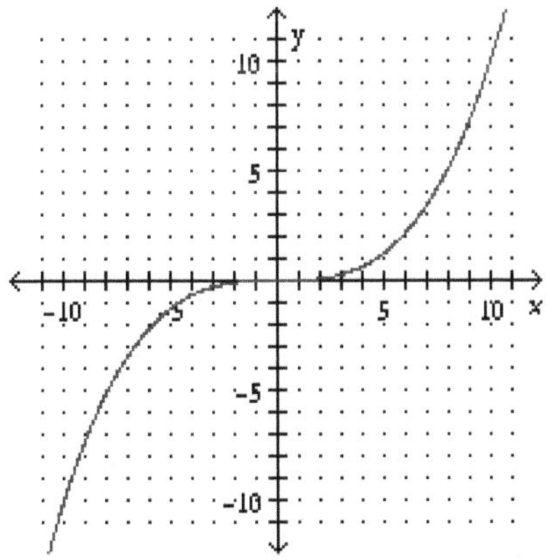

____

46.

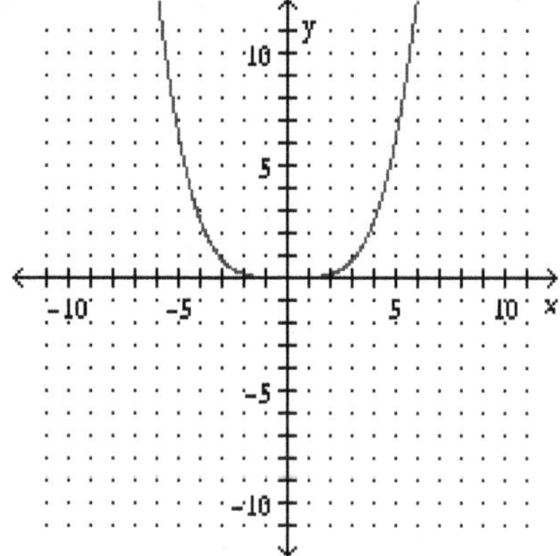
y
10
5
-10
-5
5
10
x
-5
-10

47.

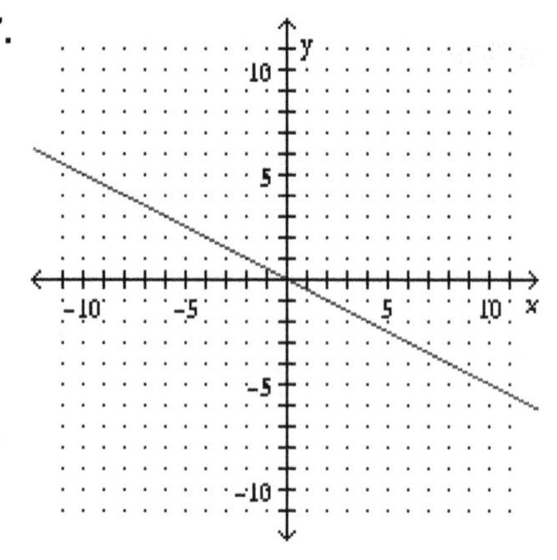
y
10
5
-10
-5
5
10
x
-5
-10

48.

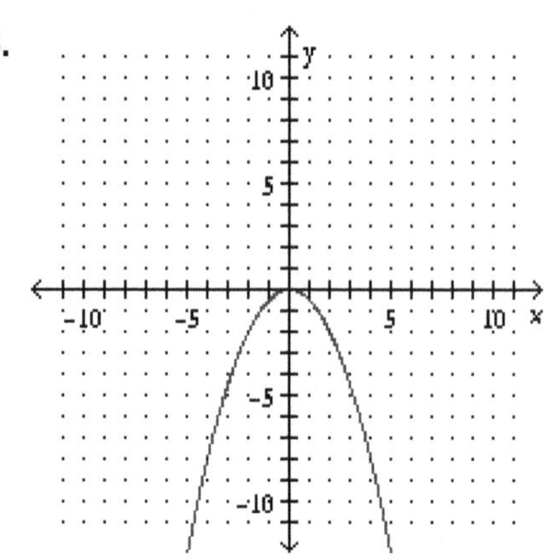
y
10
5
-10
-5
5
10
x
-5
-10

49.

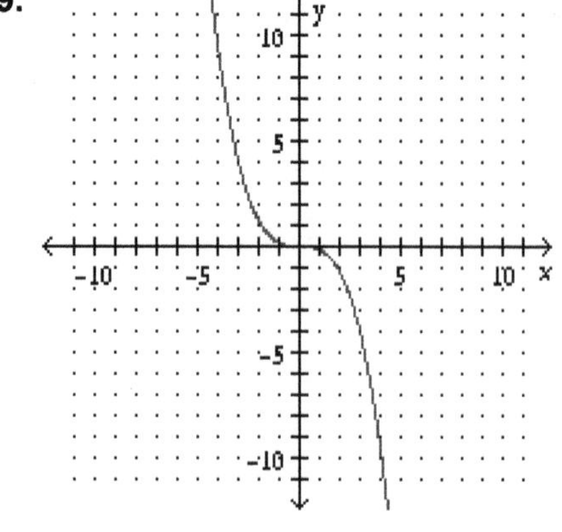
y
10
5
-10
-5
5
10
x
-5
-10

For each function graphed below, decide if the graph could be the graph of a power function. Explain your answer.

50.

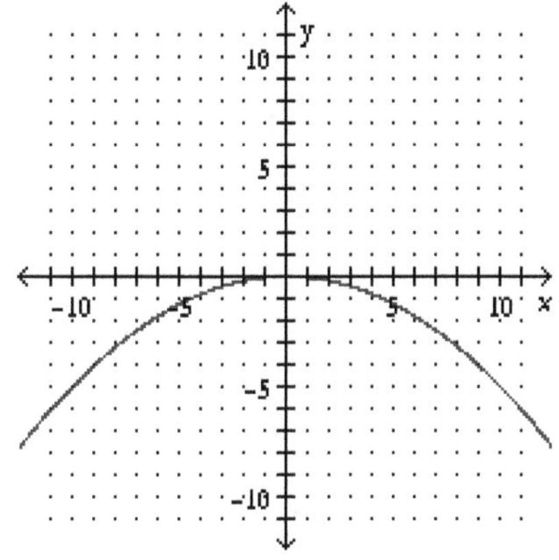

51.

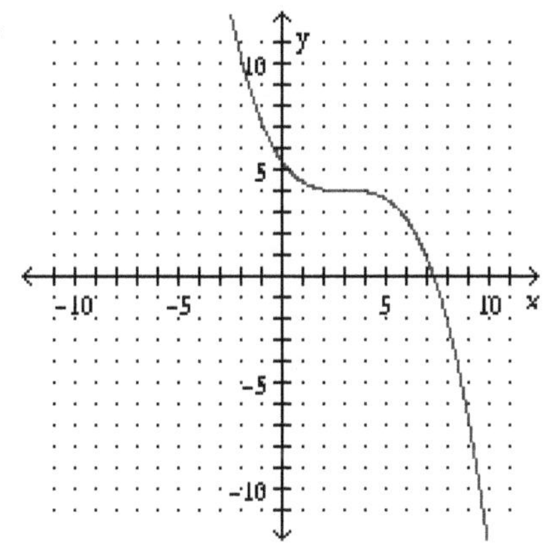

52.

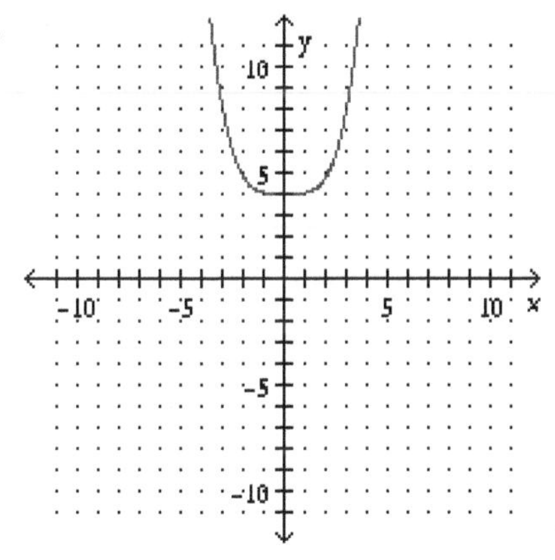

53.

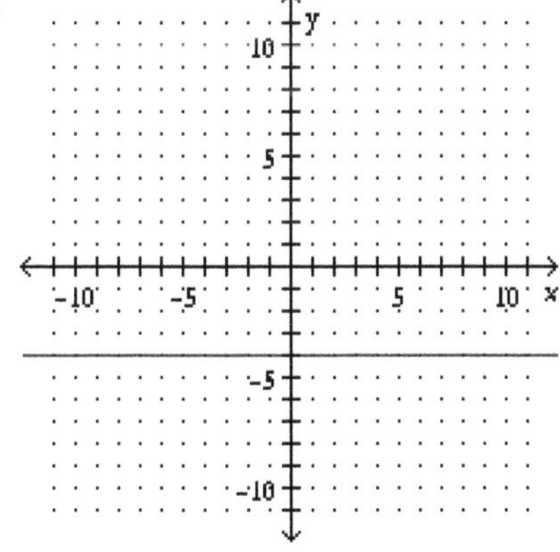

Chapter 3 Section 2: Graphs of Polynomial Functions

Practice Problems

Use the Intermediate Value Theorem to determine whether each of the following polynomial functions has real roots between the given x-values. Use only the x-values given. Explain your answer.

1. $f(x) = 2x^3 - 6x^2 + 3$ in between $x = 0$ and $x = 1$

____

2. $f(x) - 3x^3 + 5x^2 + 2$ in between $x = 1$ and $x = 3$

____

3. $f(x) = -x^2 - 3x + 4$ in between $x = -5$ and $x = 2$

____

4. $f(x) = -x^2 - 3x + 4$ in between $x = -2$ and $x = 2$

____

5. $f(x) = 2x^4 - x^2 + 3$ in between $x = -2$ and $x = 2$

____

Use the Factor Theorem to determine whether each binomial is a factor of the polynomial given. Explain your answer.

6. Is $(x-2)$ a factor of $x^3 + x^2 - 10x + 8$?

____

7. Is $(x+1)$ a factor of $x^3 + x^2 - 10x + 8$?

____

8. Is $(x-3)$ a factor of $3x^4 - 7x^3 - 3x^2 - 14x + 15$?

____

9. Is $(x+5)$ a factor of $x^3+8x^2+11x-20$?

____

10. Is $(x+3)$ a factor of $3x^4+5x^3+5x^2-2x-5$?

____

Find the zeros (and their multiplicities) of the following polynomial functions given in factored form.

11. $f(x)=(x-1)(x+2)$

____

12. $f(x)=(x+3)(x+4)^2$

____

13. $f(x)=(x+5)(x)^2(x-2)^3$

____

14. $f(x)=(x+1)^3(x-7)^4x$

____

15. $f(x)=(x+1)^2(x-2)^5$

____

16. Find possible linear factors (and their multiplicity) that would generate the polynomial expression of degree 3 described by the following graph.

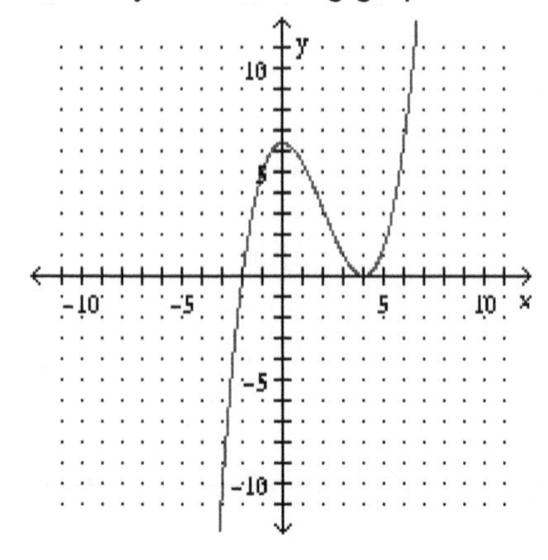

____

17. Find possible linear factors (and their multiplicity) that would generate the polynomial expression of degree 4 described by the following graph.

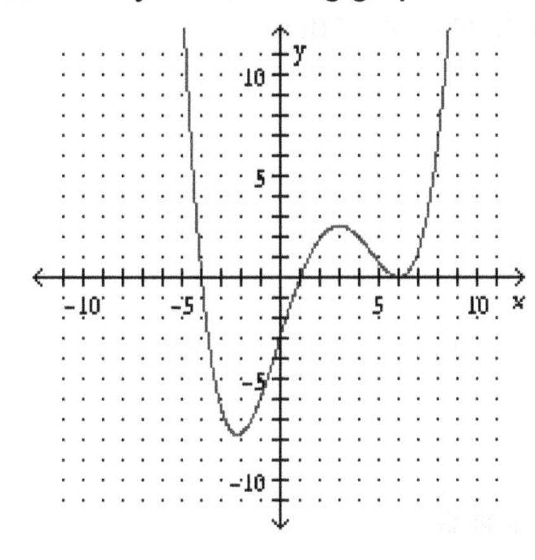

____

18. Find possible linear factors (and their multiplicity) that would generate the polynomial expression of degree 4 described by the following graph.

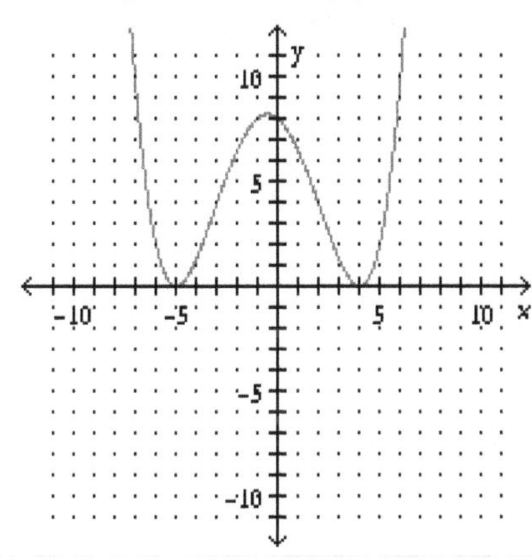

____

19. Find possible linear factors (and their multiplicity) that would generate the polynomial expression of degree 5 described by the following graph.

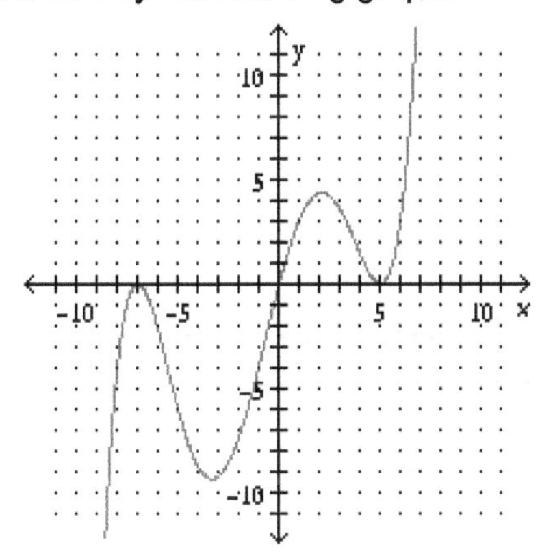

____

20. Find the algebraic expression of a polynomial of degree 2, whose graph has the following behavior:

- Crosses the x axis at $x = 1$ and at $x = -3$.
- Goes through the point $(2, 3)$.

____

21. Find the algebraic expression of a polynomial of degree 3, whose graph has the following behavior:

- Crosses the x axis at $x = 4$
- Touches the x axis at $x = -5$.
- Goes through the point $(-4, -4)$

____

22. Find the algebraic expression of a polynomial of degree 5, whose graph has the following behavior:

- Crosses the x axis at $x = 3$
- Touches the x axis at $x = 0$ and at $x = 5$.
- Goes through the point $(4, 2)$

____

23. Find the algebraic expression of a polynomial of degree 4, the graph of which has the following behavior:

- Crosses the x axis at $x = -6$ and at $x = 0$
- Touches the x axis at $x = 3$.
- Goes through the point $(2, -3)$

____

Perform each indicated division of polynomials. Give the quotient and the remainder.

24. $3x^3 + 4x^2 - 5x + 6$ divided by $x + 3$ ________________

25. $3x^4 + 5x^3 + 5x^2 - 2x - 5$ divided by $x + 1$ ________________

26. $4x^4 - 4x^2 + 2$ divided by $x - 2$ ________________

27. $2x^4 - 3x^2 - 6x + 1$ divided by $x^2 + 3$ ________________

28. $x^5 - x^4 + 4x^2 - 2x - 9$ divided by $x^2 - 3x$ ________________

A polynomial and one of its zeros is given for each problem below. Completely factor each polynomial.

29. $f(x) = x^3 - 7x^2 + 12x$; one zero is $x = 3$ ________________________

30. $f(x) = 2x^3 + 19x^2 + 45x + 18$; one zero is $x = -6$ ________________________

31. $f(x) = 6x^3 + 13x^2 - 54x - 40$; one zero is $x = -4$ ________________________

32. $f(x) = 2x^3 - 10x^2 - 24x + 120$; one zero is $x = 5$ ________________________

Use the Rational Zeros Theorem to find the set of all rational zero candidates for each polynomial function below. Then find the rational zeros of the polynomial.

33. $f(x)=x^3-4x^2+x+6$

Candidates: ____________________

Zeros: ____________________

34. $f(x)=2x^3+11x^2+17x+6$

Candidates: ____________________

Zeros: ____________________

35. $f(x)=2x^3+3x^2-17x+12$

Candidates: ____________________

Zeros: ____________________

For each graph of a polynomial function below, state if the degree of the polynomial is an even or an odd integer and give the sign of the leading coefficient. Explain your answer.

36.

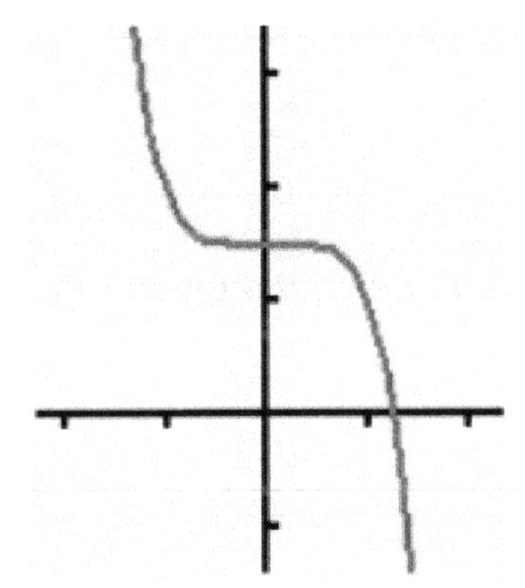

____

37.

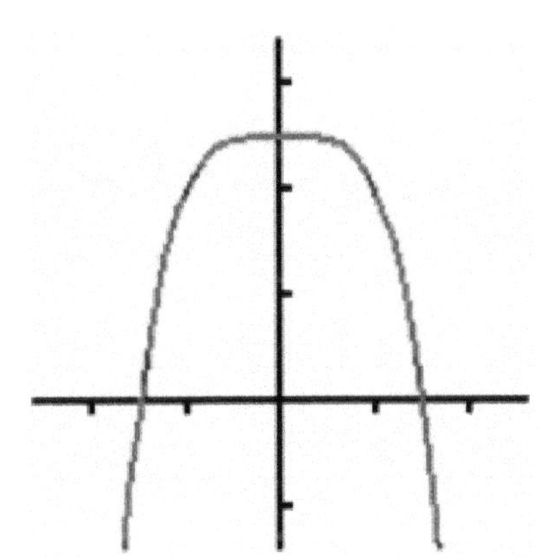

____

Find the linear factors with complex roots for the following polynomials of degree 2. Write the polynomial in factored form.

38. $f(x)=x^2+16$ ____________________

39. $f(x)=x^2+10$ ____________________

40. $f(x) = x^2 - 2x + 4$ ______________________

41. $f(x) = x^2 + 8x + 25 = 0$ ______________________

42. $f(x) = x^2 + 4x + 12 = 0$ ______________________

For each problem below, find the degree 3 polynomial with integer coefficients that has the given roots and whose graph contains the given point.

43. Roots: $\{-3, 2i, -2i\}$ through the point: $(0,12)$ ______________________

44. Roots: $\{4, \sqrt{5}\, i, -\sqrt{5}\, i\}$ through the point: $(1,36)$ ______________________

45. Roots: $\{-1, 2+i, 2-i\}$ through the point: $(2,-3)$ ______________________

46. Roots: $\left\{\frac{1}{2}, 3i, -3i\right\}$ through the point: $(3,90)$ ______________________

47. Roots: $\left\{-\frac{1}{3}, 2-3i, 2+3i\right\}$ through the point: $(1,-40)$ ______________________

Factor the following polynomial functions in the complex number system, using the given root. (Write the polynomial in factored form.)

48. $f(x) = x^3 - 4x^2 + 9x - 36$; one root is $x = 4$ ______________________

49. $f(x) = x^3 + x^2 + 7x + 7$; one root is $x = -1$ ______________________

50. $f(x) = x^3 + 7x^2 - x - 87$; one root is $x = 3$ ______________________

51. $f(x) = x^3 - 7x^2 + 16x - 10$; one root is $x = 1$ ______________________

52. $f(x) = x^3 + 8$; one root is $x = -2$ ______________________

53. $f(x) = x^3 - 2x^2 + 25x - 50$; one root is $x = 5i$ ______________________

54. $f(x) = 2x^3 + 3x^2 + 12x + 18$; one root is $x = \sqrt{6}\, i$ ______________________

55. $f(x) = x^3 + 3x^2 + 4x + 12$; one root is $x = -2i$ ______________________

56. $f(x) = 2x^3 - 5x^2 + 6x - 2$; one root is $x = 1 - i$ ______________________

57. $f(x) = x^3 - 3x + 52$; one root is $x = 2 + 3i$ ______________________

Factor the following polynomial functions in the complex number system, using the given root. (Write the polynomial in factored form.) Use your graphing calculator to help you find one of the real roots of the polynomial.

58. $f(x) = x^3 + x^2 + 16x + 16$ ______________________________

59. $f(x) = 2x^3 - 3x^2 + 14x - 21$ ______________________________

60. $f(x) = 2x^3 + x^2 + 50x + 25$ ______________________________

61. $f(x) = x^3 + 8x^2 + 9x - 58$ ______________________________

62. $f(x) = x^3 - x^2 + 8x + 60$ ______________________________

Chapter 3 Section 3: Rational Functions

Practice Problems

Identify the numbers that do not belong in the domain of the following rational functions. Explain your answer.

1. $f(x)=\frac{4}{x-2}$

2. $f(x)=\frac{5x-3}{x^2-1}$

3. $f(x)=\frac{3x^2-1}{x^2+4}$

4. $f(x)=\frac{x+2}{x^2-x-12}$

5. $f(x)=\frac{x+4}{x^2-16}$

6. $f(x)=\frac{2x+1}{2x^2+5x+2}$

7. $f(x)=\frac{3x+15}{x^2+5x}$

8. $f(x)=\frac{1}{x}+\frac{3-x}{x+2}$

Graph each of the following rational functions by studying the domain and creating a table of x-y values.

9. $f(x)=\dfrac{3}{x}$

10. $f(x)=\dfrac{2}{x-4}$

11. $f(x)=\dfrac{4x}{x+2}$

12. $f(x)=\dfrac{x-1}{2x-3}$

13. $f(x)=\dfrac{2x+5}{3x+2}$

Express the numerator and denominator of each of the following rational functions in factored form and find candidates for vertical asymptotes, holes, and zeros of each function.

14. $f(x)=\dfrac{2x+1}{2-x}$

---

15. $f(x)=\dfrac{4x-20}{3x+9}$

---

16. $f(x)=\dfrac{x^2-1}{x^2+3x+2}$

---

17. $f(x)=\dfrac{x^2-x-12}{3x^2-12}$

---

18. $f(x)=\dfrac{x^2-x-6}{x^2+6x+8}$

---

19. $f(x)=\dfrac{2x^2+x-3}{x^2-1}$

---

20. $f(x)=\dfrac{2x^2-3x-2}{2x^2+5x+2}$

---

21. $f(x) = \dfrac{3x^2 - 8x - 3}{2x^2 - 6x}$

---

22. $f(x) = \dfrac{3x^2 + 6x}{x^3 - 4x}$

---

23. $f(x) = \dfrac{2x^2 - 11x + 5}{x^3 - 3x^2 - 10x}$

The graph of each of the following functions looks like a line with a hole in it. Find the (x,y) coordinates of the hole (without graphing the function).

24. $f(x) = \dfrac{x^2 - 4x + 3}{x - 3}$ ____________________

25. $f(x) = \dfrac{x^2 - x - 20}{2x + 8}$ ____________________

26. $f(x) = \dfrac{2x^2 + 3x - 5}{1 - x}$ ____________________

*The graph of the following function looks like a line with **two** holes in it. Find the (x,y) coordinates of the holes (without graphing the function).*

27. $f(x) = \dfrac{4x^3 + 4x^2 + x}{6x^2 + 3x}$ ____________________

Describe the end behavior of each of the following rational functions. The first one is done for you as an example.

28. $f(x)=\dfrac{2x^3+4x-1}{x}$

By studying the leading terms of the numerator and denominator, we conclude that the function will behave like $g(x)=\dfrac{2x^3}{x}=2x^2$ far away from the origin of the coordinates. So at a distance the function should look like a parabola pointing upwards.

---

29. $f(x)=\dfrac{-x^3+x^2+6x}{5x-1}$

---

30. $f(x)=\dfrac{-2x^2+5x+1}{3x^2-4x+2}$

---

31. $f(x)=\dfrac{3x^2+2x-3}{4x}$

---

32. $f(x)=\dfrac{-4x^2+6x+3}{2x^2-x}$

---

33. $f(x)=\dfrac{8x^3-5x^2+3x}{-2x^3+6}$

---

34. $f(x)=\dfrac{-x^2+6x-2}{4x^4-7}$

---

35. $f(x)=\dfrac{x^2+7x}{5x^3-12x+3}$

---

36. $f(x)=\dfrac{-x^2+x-8}{2x^4-3x^3+4x}$

---

37. $f(x)=\dfrac{4x^4+5x^2-2}{-x^3+4x^2}$

Use long division to find the exact expression for the slant asymptote of each rational function below.

38. $f(x)=\dfrac{3x^2-5x+2}{x-6}$ ________________

39. $f(x)=\dfrac{-4x^2+x+1}{x+3}$ ________________

40. $f(x)=\dfrac{-3x^2+3x-8}{-x-4}$ ________________

41. $f(x)=\dfrac{2x^3+3x^2-6x}{x^2+2x}$ ________________

42. $f(x)=\dfrac{5x^3-4x^2+3}{-x^2+3x-1}$ ________________

43. $f(x)=\dfrac{-4x^3+2x^2+5}{-x^2+3x}$ ________________

44. $f(x)=\dfrac{x^4+6x^2-3}{x^3+3x^2}$ ________________

45. $f(x)=\dfrac{2x^3+3x^2-6x}{x^2+2x}$ ________________

46. $f(x)=\dfrac{x^4+1}{x^3-3x^2}$ ________________

47. $f(x)=\dfrac{x^4+3x}{x^3+1}$ ________________

Write a possible rational function with the minimum number of factors which will produce the indicated behavior. Express the numerator and denominator in factored form. You may use a graphing calculator to confirm that the function you found satisfies the requested conditions.

48.
- Hole at $x = -4$
- One vertical asymptote at $x = 0$
- One zero at $x = 3$
- Horizontal asymptote is $y = \frac{1}{2}$

____

49.
- Hole at $x = -\frac{1}{2}$
- Two vertical asymptotes at $x = -1$ and at $x = \frac{2}{5}$
- One zero at $x = 5$
- Horizontal asymptote is $y = 0$
- Goes through the point $(1, -1)$

____

50.
- Touches the x-axis at $x = 2$
- Two vertical asymptotes at $x = -1$ and at $x = \frac{2}{3}$
- Horizontal asymptote $y = \frac{5}{3}$

____

51.
- Zeros at $x = 3$ and at $x = 0$
- One vertical asymptote at $x = -2$
- Hole at $x = 1$
- Going through the point $(2, 2)$

____

52.

- Zeros at $x=-\frac{1}{2}$ and at $x=2$
- Two vertical asymptotes at $x=-3$ and at $x=0$
- Horizontal asymptote is $y=-3$

____

53. *Given the function* $f(x)=\frac{(x^2-1)(2x-5)}{(x^2+4x+3)(x-2)}$, *describe the function's behavior at each of the following* x*-values.*

a. $x=\frac{5}{2}$

____

b. $x=-3$

____

c. $x=-2$

____

d. $x=-1$

____

e. $x\rightarrow$ large positive values (end behavior towards the right)

____

54. *Given the function* $f(x)=\dfrac{(2x^2+3x-9)(x-2)}{2x^2+x-10}$*, describe the function's behavior at each of the following x-values.*

a. $x=2$

---

b. $x=-2$

---

c. $x=0$

---

d. $x=-3$

---

e. $x=-\dfrac{5}{2}$

---

f. $x\rightarrow$ large positive values

---

55. *Given the function* $f(x)=\dfrac{6x^2+x-1}{2x^3-9x^2-5x}$*, describe the function's behavior at each of the following x-values.*

a. $x=0$

---

b. $x=\dfrac{1}{3}$

---

c. $x=-\dfrac{1}{2}$

---

d. $x = 5$

---

e. $x = 1$

---

f. $x \to$ large positive values

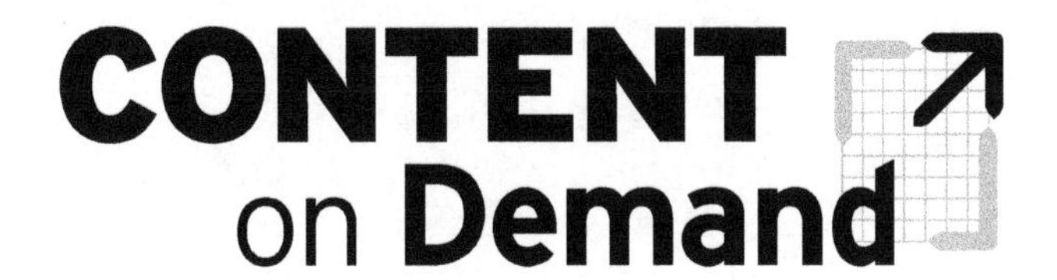

Precalculus 2nd Edition

Practice Problem Worksheets

Chapter 4: Exponential & Logarithmic Functions

Chapter 4 Section 1: Overview of Exponential Functions

Practice Problems

Approximate each value using a calculator. Express your answer rounded to three decimal places.

1. $4^{2.4}$ ________________

2. $1.2^{7.3}$ ________________

3. $0.35^{5.67}$ ________________

4. $6.38^{-2.5}$ ________________

5. $3.56^{-\pi}$ ________________

6. $6.27^{-\sqrt{2}}$ ________________

7. $\left(\frac{2}{\pi}\right)^{4.03}$ ________________

8. $\left(\frac{\pi}{5}\right)^{\sqrt{3}}$ ________________

9. $\left(\frac{2}{\sqrt{5}}\right)^{-3.46}$ ________________

10. $\left(\frac{3}{2\pi}\right)^{-\sqrt{5}}$ ________________

Graph the following exponential functions by building a table, plotting the (x, y) pairs, and joining the points with a smooth trace.

11. $f(x) = 3^x$

12. $f(x) = 2^{x-5}$

13. $f(x) = -2^{x-2}$

14. $f(x) = \left(\frac{1}{3}\right)^{x+1}$

15. $f(x) = -\left(\frac{2}{3}\right)^{\frac{x-2}{2}}$

16. $f(x) = 4 - 3^{x}$

17. $f(x) = \left(\frac{1}{2}\right)^{-x+3}$

18. $f(x) = 2^{-x+1} + 3$

19. $f(x) = 4 - 2^{\frac{x-4}{2}}$

20. $f(x) = 5 - 3^{-x-2}$

21. The graph of the exponential function $f(x) = \left(\frac{1}{4}\right)^x$ is shown below. Transform the function according to each of the following operations. Explain how the different operations will affect the graph of the transformed function, and plot it to corroborate your statements.

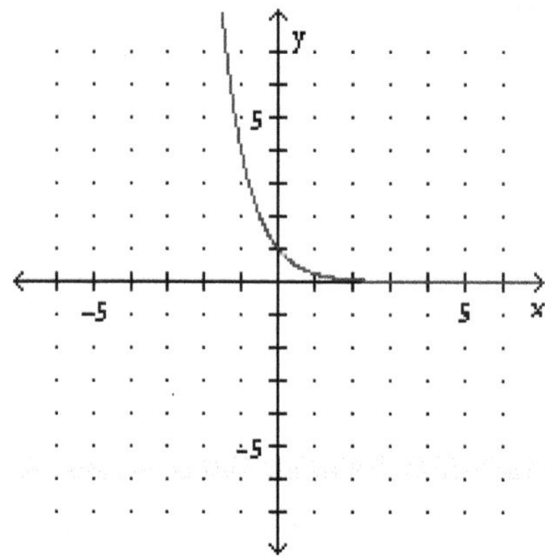

a. $f(x-5)$

b. $3 - f(x)$

c. $f(-x)$

d. $f(-x)+2$

e. $-f(-x)$

f. $4-f(-x)$

Solve for x in the following exponential equations.

22. $3^{4x-3}=9$ ____________

23. $16^{x}=\dfrac{1}{64}$ ____________

24. $25^{2x+1}=\left(\dfrac{1}{5}\right)^{x+3}$ ____________

25. $2^{5-2x}=\dfrac{1}{\sqrt{2}}$ ____________

26. $4^{2x-5}=8^{-x}$ ____________

27. $9^{\frac{2-x}{3}}=27^{2x-4}$ ____________

28. $9^{x^2}=\left(\dfrac{1}{3}\right)^{2x}$ ____________

29. $\left(\sqrt{3}\right)^{4x-6}=\left(\dfrac{1}{9}\right)^{x^2-\frac{1}{2}}$ ____________

30. $\dfrac{1}{16^{x-4}}=\left(\sqrt{8}\right)^{2x^2+8}$ ____________

31. $\dfrac{1}{125^{3x}}=5^{x^2-10}$ ____________

Write the expression of a function that models each of the following exponential growth problems.

32. A ginger root plant produces an average of 6 total tubers per season, each of which can create a new plant for the next season. If the gardener starts with 98 tubers, find a function that describes the total number of tubers produced as a function of x number of seasons.

____

33. The number of aphids on a rose plant doubles in a week. If the initial number of aphids is 25, write an exponential function that describes the number of aphids as a function of x number of weeks.

____

34. The population of mice at the harbor grows at a rate of 32% per month. If the initial population is 250 mice, write a function that describes the population as a function of x months.

____

35. The population of a small town is growing at a rate of 8.5% per year. If the initial population is 265,000 inhabitants, write a function that describes the population as a function of x number of years.

____

36. The number of millionaires in a developing country is growing at a rate of 4.7 percent per year. If the initial number of millionaires is 1358, write a function that describes the number of millionaires as a function of x number of years.

____

37. A farmer started with a pair of rabbits at the beginning of the year. Six months later the rabbit population has grown to 22 rabbits total. Considering that the population will continue increasing at this rate, write an exponential function that describes it as a function of time (x) in months, starting with the original couple.

____

Write the expression of a function that models each of the following exponential decay problems.

38. The falcon population in a wilderness area has decreased from 270 to 216 in one year. If this rate of decrease is maintained, find an exponential expression for the number of falcons as a function of time (x) in years, starting with 270 individuals.

____

39. The percent of unemployment in a country with expanding industry has decreased exponentially from 38% to 27% in one year. If this trend continues, find an expression for the percent of unemployment as a function of x years.

____

40. Water is recycled through a filtering device for purification. In every pass, 18% of the impurities are removed. If the water started with 157grams of impurities, write an exponential expression for the amount of impurities (in grams) per (x) number of passes through the filter.

____

41. Antibiotic in a patient's blood is excreted at a rate of 12.5% per hour. Find an expression for the amount of antibiotic present (in milligrams) after x hours if the starting dose was 500mg.

____

42. The price of a standard desktop computer has come down exponentially from $2400 to $1000 in five years. If this trend is kept, write an exponential expression for the computer's price as a function of (x) years starting at $2400.

____

For each problem below, use the compound interest formula to calculate the balance in a deposit account with the stated conditions.

43. Principal: $5000; Annual rate: 6.0%; Compounded: semiannually and after 3 years.

____

44. Principal: $10,500; Annual rate: 5.6%; Compounded: monthly and after 5 years.

____

45. Principal: $3250; Annual rate: 3.7%; Compounded: annually and after 4 years.

____

46. Principal: $5400; Annual rate: 5.1%; Compounded: semiannually and after 5.5 years.

____

47. Principal: $900; Annual rate: 4.5%; Compounded: daily and after 2 years.

____

48. Principal: $1500; Annual rate: 6.3%; Compounded: weekly and after 3.5 years.

____

49. Principal: $14,400; Annual rate: 4.7%; Compounded: quarterly and after 4.5 years.

____

Chapter 4 Section 2: Exponential Functions of Base e

Practice Problems

*Approximate each value using a calculator. Express your answer rounded to **four** decimal places.*

1. $e^{2.5}$ ________________

2. e^{π} ________________

3. $e^{-\frac{\pi}{2}}$ ________________

4. $\dfrac{1}{e^{\frac{\sqrt{5}}{2}}}$ ________________

5. $\dfrac{4}{1+e^{-\sqrt{2}}}$ ________________

6. $62\ e^{-2\pi}$ ________________

7. $150\ e^{-0.124\ \pi}$ ________________

8. $\dfrac{60}{1+25\ e^{-0.45\sqrt{3}}}$ ________________

9. $-e^{3.42\sqrt{0.5}}$ ________________

10. $\dfrac{67}{1-e^{-0.03\sqrt{7}}}$ ________________

11. The functions $f(x)=e^{-x}+2$, $g(x)=2^{-x}+2$ and $h(x)=3^{-x}+2$ are plotted below. Identify each of them and explain the reasons for your choice.

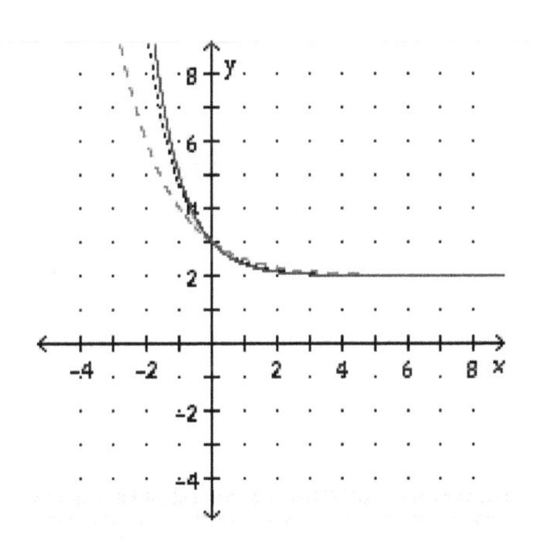

12. The functions $f(x) = -e^{x-4}$, $g(x) = -2^{x-4}$ and $h(x) = -3^{x-4}$ are plotted below. Identify each of them and explain the reasons for your choice.

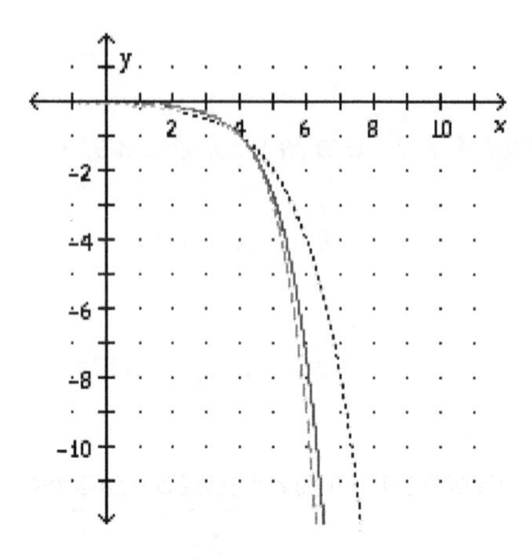

Solve for x.

13. $e^{-x} = e^{x+1}$ ______________

14. $\dfrac{1}{e^3} = e^{2x-1}$ ______________

15. $e^{3x-5} = 1$ ______________

16. $5e^{2x+6} = 6e^{2x+6} - 1$ ______________

17. $\sqrt{e} = \left(e^3\right)^{2x+1}$ ______________

18. $\dfrac{1}{e^{2x}} = \sqrt[3]{e^{4x}}$ ______________

19. $\dfrac{1}{e^{-x+3}} = \sqrt[3]{e^{5x}}$ ______________

Write an exponential expression that models each event and use the model you created to answer the questions.

20. A population of some protozoa (unicellular microscopic organisms), grows at a **continuous** percent rate of 342% per hour. If the initial protozoan count was 400:

a) What will the protozoan count be in 1 hour?
b) What will it be in 1hour 30 minutes?
c) What will it be in 2 hours and 45 minutes?
d) What will it be in 20 minutes?

____

21. A population of mice grows at a continuous rate of 8% per month. If the initial population is 350 mice today, how many individuals will there be in:

a) 4 months?
b) 28 weeks?
c) 42 weeks?
d) 1 year?

____

22. The price of a car decreases with time at a continuous 37.2% rate per year. If the car's price when new was $36,450, estimate its price:

a) 3 months after the purchase
b) Half a year after the purchase
c) 18 months after the purchase
d) 2 years after the purchase

____

23. The population of a species of snake in a rain forest is decreasing at a continuous 2.5% per year. If today there are 1500 snakes in the rain forest, estimate the snake population in:

a) 21 months
b) 2.5 years
c) 7 years
d) 20 years

____

24. An isotope used for medical tracing decays at a continuous 1.1% percent per minute. If a patient was given 6300 milligrams for a scanning procedure, how much of the isotope will be present in the patient after:

a) 20 minutes?
b) 3 hours?
c) 8 hours ?
d) one day?

____

25. Compare the total amount accumulated in a savings account at 4.2% annually compounded interest to that of a 4.2% rate but **continuously** compounded, when $7500 is deposited for 8 years.

____

26. Which annually compounded interest rate "*r*" will produce the same total amount as that of a savings account with a 5.2% **continuously** compounded rate in a:

a) One year period?
b) Two year period?
c) Five year period?

____

27. The bison population in a national preserve responds to the following logistic function:

$$P(t) = \frac{350}{1+6\,e^{-0.181\,t}}$$

where "t" represents the number of years after the initial bison population was introduced into the preserve. Find:

a) What was the starting population?
b) What was the bison population 10 years later?
c) What will the number of individuals be as the bison population approaches larger values of "t"?

____

28. The number of people infected with a viral disease as it spreads over the population of a small town is described by the following logistic function: $P(t) = \dfrac{7850}{1+3924\,e^{-2.41\,t}}$

where "t" represents the number of weeks after the first cases where detected. Find:

a) How many people where infected to start with?
b) What was the number of people infected after 2 weeks?
c) After 1 month?
d) What will be the number of individuals that will get the disease in the long run?

____

29. A fossil is analyzed for carbon-14 content in order to determine its age. The results from the lab come with the statement that the fossil is supposed to be about 6000 years old. What was the approximate percent of carbon-14 the lab scientists found in the fossil when they analyzed it?

____

30. A potato that was on the counter at ambient temperature (57°F) is placed in a 475°F oven. Assuming that the heating of the potato responds to Newton's law of heating and cooling, with a characteristic parameter k of 0.124 (per minute), write a functional expression for the potato's temperature as a function of time in the oven (in minutes), and find the potato's temperature after:

a) 10 minutes
b) 25 minutes
c) 1 hour

____

31. A piece of pottery is removed from a 625°F oven and placed on a rack to cool at ambient temperature (62°F). Assuming that the cooling of the pottery responds to Newton's law of heating and cooling with a characteristic parameter k of 0.097 (per minute), write a functional expression for the pottery's temperature as a function of time on the cooling rack (in minutes), and find its temperature after:

a) 20 minutes
b) 45 minutes
c) 1 hour and 10 minutes

____

Appendix: Fitting Exponential Functions to Data

Practice Problems

Use the regression option ("Regr") of Calculator on Demand to find the exponential function that best fits the data shown in each table, and give the coefficient of determination.

1.

x	0	3	5	7	10
y	10	22	56	86	127

2.

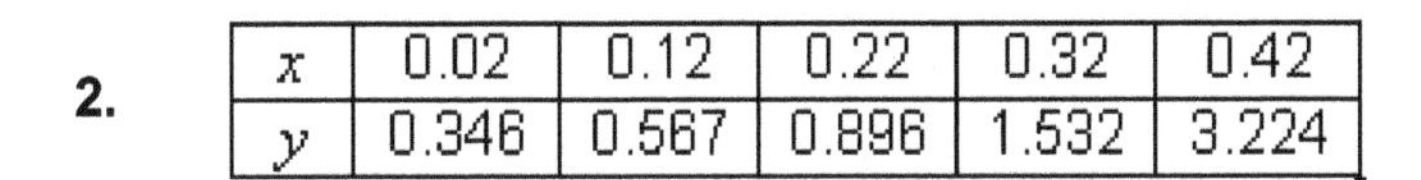

x	0.02	0.12	0.22	0.32	0.42
y	0.346	0.567	0.896	1.532	3.224

3.

x	57.2	82.1	95.7	111.8	135.2
y	134	98	75	63	44

4.

x	0.03	0.15	0.43	0.78	0.91
y	3.21	1.24	0.65	0.41	0.25

5. The following table shows the population (in thousands of inhabitants) of a town for different years. Use the regression option in your graphing calculator to find the best exponential function that fits the data (use the year 1990 as year "zero"); use the function you find to predict this town's population in the year 2020.

Year	1990	1992	1994	1996	1998	2000
Population	38.6	51.2	73.1	89.6	114.3	132.0

6. A scientist in a nuclear physics lab is measuring the amount a radioactive substance in an initial sample of 200 grams as a function of time. The following table indicates the different amounts of the radioactive substance left at different times. Find the exponential function that best fits the given data. Based on the expression found, predict the amount of radioactive substance left after 48 hours.

Hours	0	2	4	8	10	12	14
Rad.material (grams)	200	175	150	120	100	90	75

7. Oak trees start producing acorns once they are approximately 25 years old. The following table indicates the average number of acorns that an oak tree produces at different ages:

Age (years)	25	30	40	50	60	70
# of acorns	20	30	50	120	300	700

Use the "Regression" tool in your calculator to find which of the following "Regression Models" reproduces the data best: "Linear," "Quadratic," "Exponential," or "Power". Give the best functional expression and explain the reasons for your selection.

8. The average "per capita" personal income in the state of Indiana for different years is given in the following table:

Year	1969	1974	1979	1984	1989	1994	1999
Income ($)	2,877	4,170	7,500	10,569	12,747	15,954	20,635

Use the "Regression" tool in your calculator to find which of the following "Regression Models" reproduces the data best: "Linear," "Quadratic," or "Exponential". Take the year 1969 as year "zero". Give the best functional expression and explain the reasons for your selection.

Use the best function obtained to predict the average "per capita" personal income in the state in the year 2009.

Chapter 4 Section 3: Logarithms

Practice Problems

Solve for x in each logarithmic equation below (give the exact answer). If the equation does not have an exact decimal solution, give the answer rounded to 5 decimals.

1. $\log_4\left(\frac{x}{2}\right)=3$ ____________

2. $\log_5(3x-7)=0.2$ ____________

3. $\frac{1}{2}\log_3(2x)=-1.5$ ____________

4. $3\log_{6.2}(3x+1)=-5$ ____________

5. $4.5\ \log_{5.1}(2x-4)=-5$ ____________

*Find each logarithm **without** the use of a calculator.*

6. $\log\left(\sqrt{1000}\right)$ ____________

7. $\ln\left(\frac{1}{e^{-3}}\right)$ ____________

8. $\log_5\left(\frac{1}{125}\right)$ ____________

9. $\log_3\left(\frac{27}{\sqrt{3}}\right)$ ____________

10. $\log_2\left(\frac{\sqrt[3]{2^4}}{8}\right)$ ____________

Solve for x in each exponential equation below (give the exact answer). If the equation does not have an exact decimal solution, give the answer rounded to 5 decimals.

11. $\left(\frac{1}{2}\right)^{-x} = 4$ ________________

12. $2^{3x-1} = \frac{1}{16}$ ________________

13. $9^{x-2} = \frac{1}{27}$ ________________

14. $2 + 10^{-x+1} = 7.2$ ________________

15. $5.3 - \frac{7}{10^{2x+3}} = -6.1$ ________________

16. $1.3 - 10^{-4x+2} = 5.2$ ________________

17. $5.2 + \frac{1}{e^{-2x}} = 7.3$ ________________

18. $4e^{3x-2} - 3 = 5.8$ ________________

19. $50 = \frac{150}{1 + 3 \cdot 10^{-2x}}$ ________________

20. $70 = \frac{35}{1 - 4\, e^{-2x}}$ ________________

Use your calculator to find each logarithm. Round your answer to five decimals.

21. $\log\left(3.2\ 10^{2}\right)$ ________________

22. $\ln\left(5.7\ e^{-4}\right)$ ________________

23. $\log\left(10^{-3}\ e^{7}\right)$ ________________

24. $\ln\left(\frac{3\pi}{e^{2}}\right)$ ________________

25. $\log\left(\frac{\sqrt{\pi}}{10^{-3}}\right)$ ________________

Use the change of base formula to find each logarithm. Round your answers to five decimals.

26. $\log_5(3.7)$ ____________

27. $\log_{\frac{1}{2}}(45.7)$ ____________

28. $\log_{\pi}(32.6)$ ____________

29. $\log_{\frac{e}{4}}(0.59)$ ____________

30. $\log_{\frac{3\pi}{2}}\left(e^3\right)$ ____________

Expand each logarithm as a sum or difference of logarithms of linear expressions.

31. $\log\left(\frac{x-4}{x^2}\right)$ ____________________

32. $\log\left(6x^2-x-1\right)$ ____________________

33. $\log\left(\frac{\sqrt{x-3}}{x^2-1}\right)$ ____________________

34. $\ln\left(\frac{x^2+4x+4}{\sqrt[3]{x+1}}\right)$ ____________________

35. $\ln\left(\sqrt[4]{\frac{x^2-9}{x+5}}\right)$ ____________________

Write as a single logarithm.

36. $\log(x-1)+\log(x+2)-\log(3x-4)$ ____________________

37. $3\ \ln(x+6)-2\ \ln(x+1)$ ____________________

38. $\frac{2}{3}\left[\log(x+2)+\log(x-2)\right]$ ____________________

39. $\frac{2}{3}\left[\log(x-3)+\log(x)\right]-\frac{1}{3}\log(x+4)$ ____________________

40. $4\ln(x)-2\ln(2x+1)-\frac{1}{2}\ln(x-3)$ ____________________

Write each equation without logarithms.

41. $\ln(x) + 2\ln(y) = 3\ln(4)$ ______________

42. $\log(x) + \frac{1}{2}\log(y) = 1$ ______________

43. $2\ln(x-1) - 2\ln(y) = \ln(7)$ ______________

44. $3\log(3x-1) = 2\log(y)$ ______________

45. $2\ln(x+3) = y\ \ln(4) + 2$ ______________

Solve each exponential equation.

46. $3^{x-4} = 4.5$ ______________

47. $2.5^{2x-1} = 3.6$ ______________

48. $2.3 = 4.5\ 8^{-3x+2}$ ______________

49. $5.2 = \dfrac{1.6}{4.1^{3-x}}$ ______________

50. $3.1^{-2x} = 0.3^{x+1}$ ______________

Knowing that $\log_b(2) = 0.756$ *and* $\log_b(3) = 1.199$, *use the properties of logarithms to find each of the following.*

51. $\log_b(6)$ ______________

52. $\log_b\left(\frac{9}{8}\right)$ ______________

53. $\log_b\left(\sqrt{24}\right)$ ______________

54. $\log_b\left(\frac{\sqrt[3]{9}}{4}\right)$ ______________

55. $\log_b\left(72\ \sqrt[3]{18}\right)$ ______________

Knowing that $\log_b(A) = -0.348$ *and* $\log_b(C) = 1.252$*, use the properties of logarithms to find each of the following.*

56. $\log_b\left(A \cdot C^2\right)$ ____________

57. $\log_b\left(\dfrac{C^2}{A^3}\right)$ ____________

58. $\log_b\left(\sqrt{A^3 \cdot C}\right)$ ____________

59. $\log_b\left(\dfrac{A^3}{\sqrt{C}}\right)$ ____________

60. $\log_b\left(\dfrac{A^2\sqrt[4]{C}}{C^2}\right) - 2$ ____________

Solve each logarithmic equation.

61. $\ln\left(x^2 - 6x + 10\right) = \ln(2x - 5)$ ____________

62. $\log(3x - 6) - \log(x) = \log(x + 8)$ ____________

63. $\ln\left(x^2 - 9x + 14\right) - \ln(x - 2) = 0$ ____________

64. $\log(2x) - \log(x + 4) = 0$ ____________

65. $\log(6) - \log(4 - 3x) = \log(3)$ ____________

66. $\log(15) - \log(x + 4) = \log(x - 4) - \log(x)$ ____________

67. $\ln(5x - 10) - \log(3x - 8) = \log(x)$ ____________

68. $\ln(2x) - \ln(x + 4) = \ln(3) - \ln(x - 1)$ ____________

69. $\ln(x) - \ln(x - 4) - \ln(x + 4) + \ln(6) = 0$ ____________

70. $\log(x - 3) - \log(x) = \log\left(\dfrac{4}{7}\right)$ ____________

71. $\log_2(-3x - 2) = 1$ ____________

72. $\log_3(x + 8) + \log_3(x) = 2$ ____________

73. $\log_3(x + 2) + \log_3(x) = 1$ ____________

74. $\log_4(x + 12) = 3 - \log_4(x)$ ____________

75. $\log_3(x - 8) = 2 - \log_3(x)$ ____________

Chapter 4 Section 4: Applications of Logarithms and Exponentials

Practice Problems

1. A table of population for various cities is given below. Complete the table by finding the missing values. Assume a **continuous** rate of growth in all cases.

City	Initial Population (in thousands)	Growth rate per year (in %)	Years elapsed	Current Population (in thousands)
A	125	2.6	32	?
B	?	8.5	24	376
C	346	?	17	472
D	515	6.7	?	861

a. City A

____

b. City B

____

c. City C

____

d. City D

____

2. The current price of a piece of jewelry is $1140. The jewelry's value has shown a **continuous** growth rate of 4.6% per year. What was the jewelry's value 10 years ago?

____

3. A coin collector bought a set of old coins which shows a **continuous** growth rate of 3.9% per year. The investor paid $865 for the coin set.

a. Give a function that projects the set's value as a function of time (in years) from the purchase.

____

b. Find how long the investor has to wait in order to double the value of the original purchase.

____

4. A $2000 investment in a continuously compounded savings account yielded an interest of $345 after 3 years. Find the annual interest rate of the account.

____

5. An incomplete table of radioactive measurements from a nuclear laboratory is given below. The different quantities of the samples used are shown, as well as some of their "half lives," continuous decay rate, elapsed time of the experiments, and final amount of the radioactive substance in the sample. Complete the table, using the continuous decay expression: $N(t) = N_0\ e^{-k \cdot t}$.

Substance	Initial weight (in grams)	Elapsed time (hours)	Final weight (in grams)	Half life (hour)	Decay rate (%/hour)
A	50	17	?	6.5	?
B	35	10 and 1/2	?	?	14.2
C	?	16	12.6	13.5	?
D	73	20	46.2	?	?
E	43	?	19.5	5.8	?

a. Substance A

____

b. Substance B

____

c. Substance C

____

d. Substance D

____

e. Substance E

____

6. The results of carbon-14 dating on a petrified plant shows that the plant died approximately 10750 years ago. What was the percent of carbon-14 present in the fossil at the time the dating was performed?

____

7. Wooden artifacts found in at an excavation site contain 63.2% of the original amount of C14. How old are the artifacts?

____

8. In the middle of autumn, the number of leaves left on a tree are modeled by an exponential decay function with a continuous rate. The tree started with 3000 leaves and in four days only 122 leaves were left.

a. Find the exponential expression with base "*e*" that describes the number of leaves left.

____

b. How many days will it take until there is only one leaf left on the tree?

____

9. The number of orangutans in the rain forests of Borneo has been decreasing at a continuous rate from about 25,000 individuals in the year 1980 to about 1800 twenty years later (in 2000). If this rate of decrease continues, when will the orangutan population be virtually extinct (only one individual left)?

____

10. In a typing class, the function that models the number of words per minute that a student types as a function of days of training is given by the following "learning curve":

$$n(t) = 60 - 50\ e^{-0.103\ t}$$ where "*t*" is given in days.

a. How many words per minute is a starting student supposed to be able to type?

____

b. How many words per minute would a student be able to type after one week of training?

____

c. What is the maximum number of words per minute that a student is expected to type?

____

11. The function that describes the charging of a capacitor (in Volts) as a function of time (*t*), is given by:

$$V(t) = 450\ \left(1 - e^{-0.701\ t}\right)$$ where "*t*" is given in seconds.

a. What is the starting voltage in the capacitor?

____

b. Which voltage does the capacitor reach after charging for 4 seconds?

____

c. What is the maximum voltage that the capacitor can reach?

____

12. The function that describes the battery charging (in Volts) of a of a cell phone is given by:

$$V(t) = 12.5 - 3.5\ e^{-0.013\ t}$$ where "*t*" is given in minutes.

a. What is the starting voltage in the cell phone battery?

____

b. Which voltage does the battery reach after charging for 30 minutes?

____

c. What is the maximum voltage that the battery can reach?

____

13. The score obtained by students in an algebra course (from 1 to 100) depends on the number of hours devoted to studying the subject and it is modeled by the following function:

$$S(t) = 100 - 90\ e^{-0.0085\ t}$$ where "*t*" is the number of hours the student studied.

a. How many hours minimum should the student study the subject to obtain a "C" (70%)?

____

b. How many hours minimum should the student study the subject to obtain a "B" (80%)?

____

c. How many hours minimum should the student study the subject to obtain an "A" (90%)?

____

14. The spreading of an avian flu virus in California and Texas can be modeled by the function:

$$P(t)=\frac{100}{1+99\ e^{-0.0787\ t}}$$

where $P(t)$ is the percent of chicken that contracted the virus and "t" is measured in hours.

a. What percent of chicken were infected with the virus when the flu was first detected?

____

b. What percent had contracted the virus 2 days later?

____

c. What percent had contracted the virus 3 days later?

____

15. Due to the news coverage during political campaigns, the spread of rumors and news among the electorate can be modeled by the following function:

$$P(t)=\frac{100}{1+99\ e^{-0.8545\ t}}$$ where $P(t)$ is the percent of the electorate who has heard a rumor "t" days after the rumor was originated. If a political party plans to spread a rumor about their opponent and they want 80% of the electorate to know it by election day, how many days before the election should the rumor be started?

____

16. A car depreciates with time according to the following function:

$V(t)=24400\ e^{-k\cdot t}+1600$; where $V(t)$ is given in dollars, and "t" in years.

The depreciation rate constant "k" is not given, but we know that the car's value one year after purchase is \$20,450.

a. Find the purchase value of the car.

____

b. Find the value of "k".

____

c. In how many years will the car's value be reduced to one half of its original price?

____

17. A cup of tea served at 210°F is placed on a table in a room at 52 °F. Five minutes later, the temperature of the tea is 129 °F. How much longer will it take to reach 80 °F? Assume that Newton's law of cooling describes the cooling process.

____

18. A piece of ham from the refrigerator (at 45 °F) is placed in a 325 °F oven to cook. If in 10 minutes the ham's temperature reaches 65 °F, how many minutes until the ham reaches 180 °F?

____

19. The following functions describe the growth of different populations of animals in a preserve as a function of time (in years) starting in 1990 (as year zero).

Population A: $A(t) = 0.4\,e^{0.235\,t}$ (in thousands of individuals)

Population B: $B(t) = \dfrac{3.2}{1 + 4.5\,e^{-0.34\,t}}$ (in thousands of individuals)

Population C: $C(t) = 3.1 - 2.8\,e^{-0.51\,t}$

Use your graphing calculator to find the following.

a. When will population A be equal to population B? What is the number of individuals in each population when this happens?

____

b. When will population A be equal to double population C? What is the number of individuals in each population when this happens?

____

c. When will population B plus population C equal population A? What is the number of individuals in each population when this happens?

____

CONTENT on Demand

Precalculus 2nd Edition

Practice Problem Worksheets

Chapter 5: Trigonometric Functions

Practice Problems

Find the degree measure of each standard angle depicted below.

1.

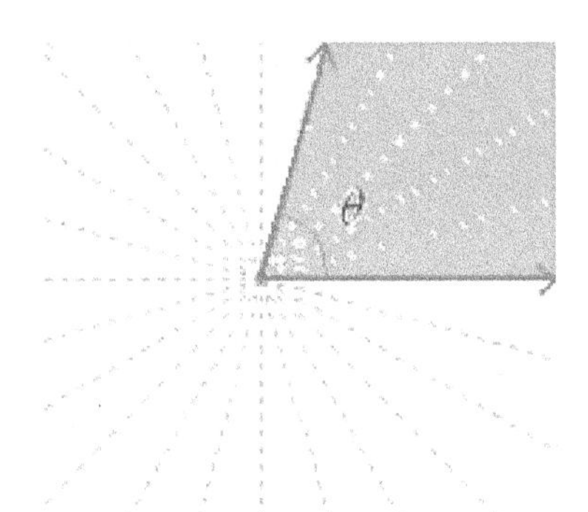

2.

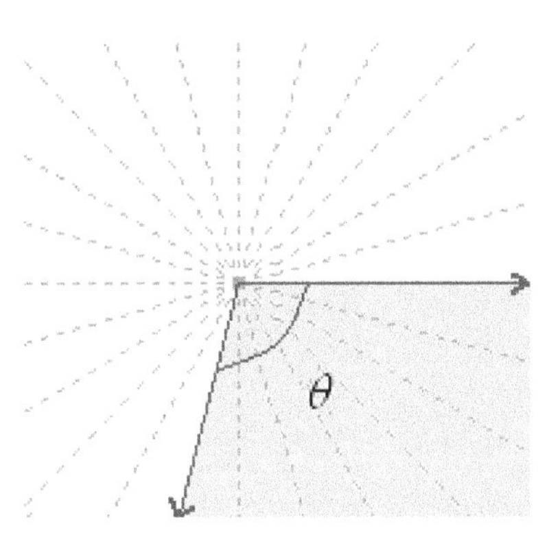

3.

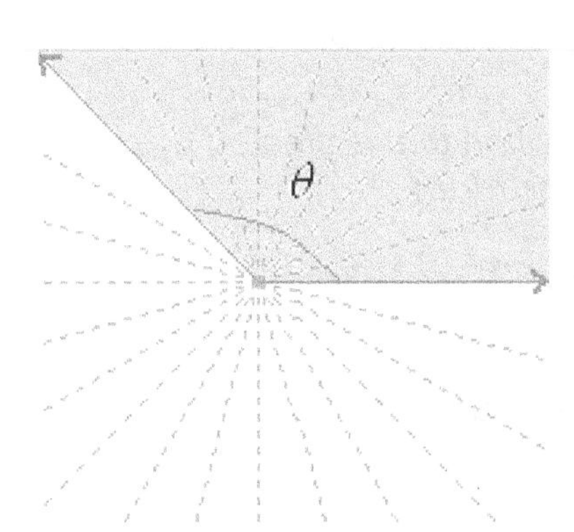

4.

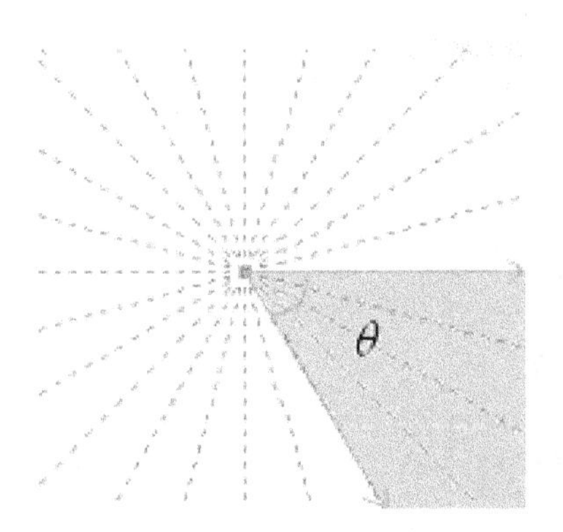

Graph each specified angle on the grids provided.

5. $\theta = -30^\circ$

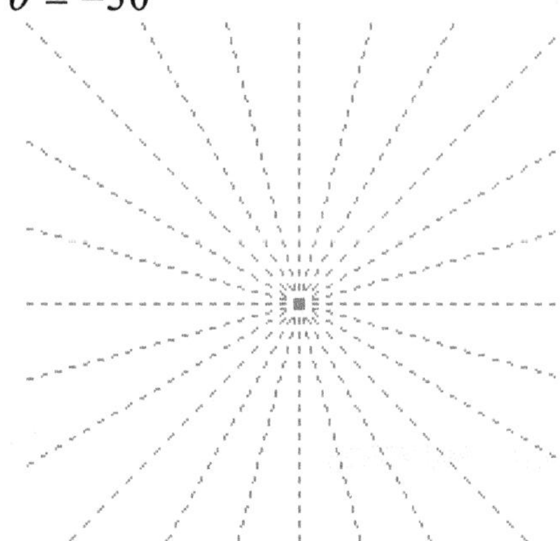

6. $\theta = 270^\circ$

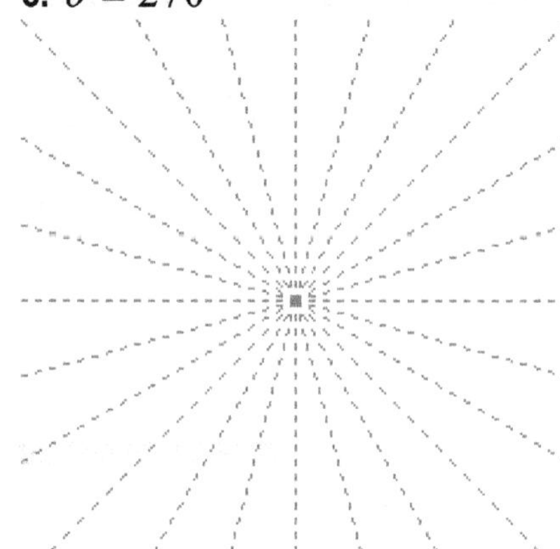

7. $\theta = -75^\circ$

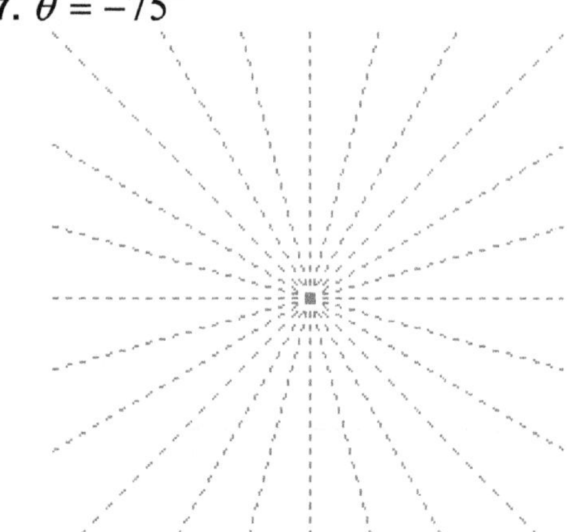

8. $\theta = 240^\circ$

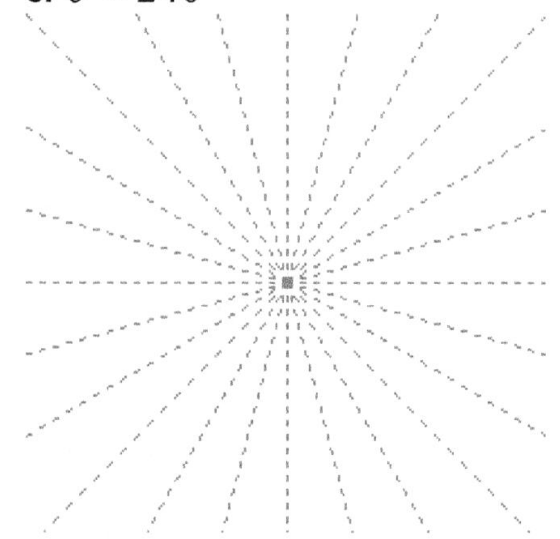

Find the radian measure of the angle in standard position that is created by rotating the terminal side as described (write your answer in terms of π).

9. Half a circle ________

10. $\frac{1}{4}$ of a circle ________

11. $\frac{2}{3}$ of a circle ________

12. $\frac{3}{8}$ of a circle ________

13. $\frac{1}{8}$ of a circle ________

Find the radian measure of an angle in standard position, within $[0, 2\pi]$, that is coterminal to each angle given.

14. 8π ________

15. 5π ________

16. $\frac{8\pi}{3}$ ________

17. $\frac{7\pi}{2}$ ________

18. $\frac{17\pi}{4}$ ________

19. $-\frac{\pi}{4}$ ________

20. $-\frac{\pi}{2}$ ________

21. $-\frac{7\pi}{5}$ ________

22. $-\frac{9\pi}{4}$ ________

Convert the angles given in radians below into degrees.

23. $\frac{\pi}{3}$ __________ **24.** $\frac{\pi}{5}$ __________

25. $\frac{\pi}{10}$ __________ **26.** $\frac{3\pi}{10}$ __________

27. $-\frac{\pi}{6}$ __________ **28.** $-\frac{5\pi}{3}$ __________

29. $-\frac{2\pi}{5}$ __________ **30.** $-\frac{7\pi}{4}$ __________

31. $\frac{19\pi}{2}$ __________ **32.** $-\frac{5\pi}{12}$ __________

Convert the angles given in degrees below into radians (write your answer in terms of π).

33. 144° __________ **34.** 630° __________

35. 405° __________ **36.** 450° __________

37. -252° __________ **38.** -20° __________

39. 80° __________ **40.** -72° __________

41. -420° __________ **42.** -225° __________

The arc described by the tip of a rod of a given length that rotates around one of its ends is given below. Give the measure of each subtended angle, in radians.

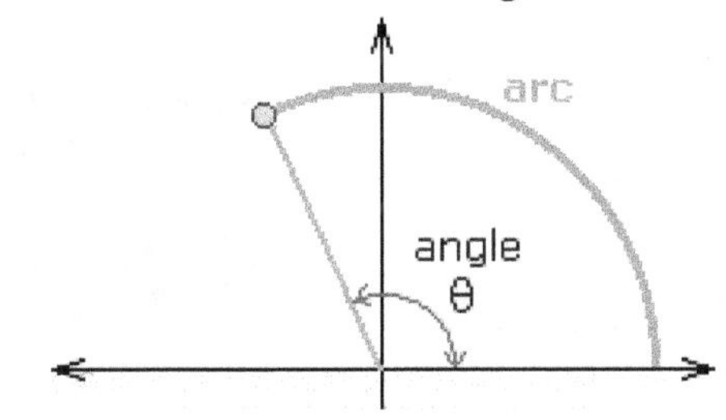

43. arc = 35 in
length of rod = 15 in __________

44. arc = 50 cm
length of rod = 20 cm __________

45. arc = 46 in
length of rod = 12 in __________

46. arc = 36 m
length of rod = 8 m __________

47. arc = 45 ft
length of rod = 12 ft __________

The latitude of a location on Earth is the angle between the location and the plane that goes through the equator, having the angle's vertex at the center of the Earth. Assuming the Earth to be a perfect sphere of radius 6400 km, find the distance (in km) on the Earth's surface between each of the following cities and the North Pole.

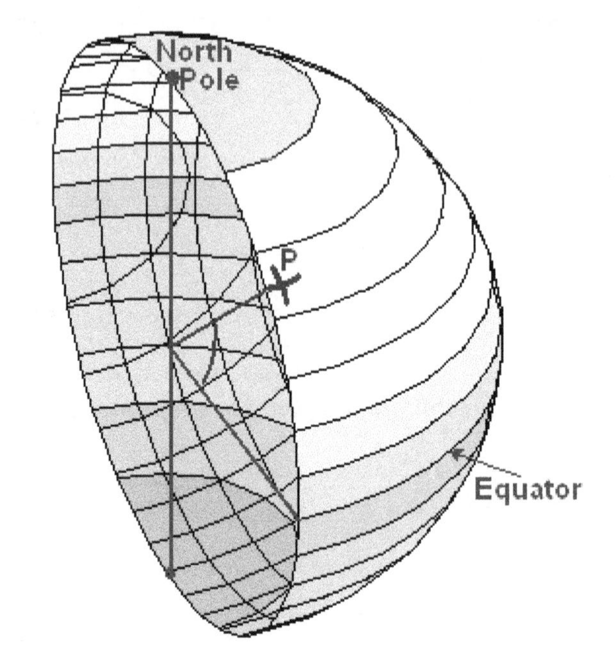

48. Los Angeles, USA
Latitude 34° north __________

49. Leon, Mexico
Latitude 21° north __________

50. Edinburgh, Scotland
Latitude 57° north __________

51. Recife, Brazil
Latitude 3° south __________

52. Auckland, New Zealand
Latitude 37° south __________

A wheel rotates at different rates, given below in revolutions per minute (rpm). For each case below, find the angular speed of the wheel (in radians per minute) and the linear speed of a point "P" (rounded to the nearest whole number) in the given units of distance per minute.

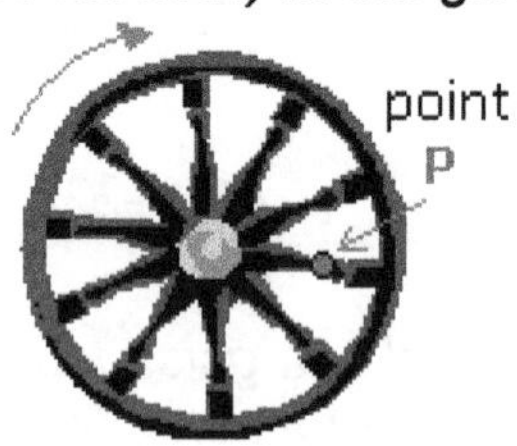

53. rate = 25 rpm
"P" = 1 ft from center __________

54. rate = 100 rpm
"P" = 6 in from center __________

55. rate = 35 rpm
"P" = 25 cm from center __________

56. rate = 50 rpm
"P" = 0.5 m from center __________

Chapter 5 Section 2: Trigonometric Functions from Circular Motion

Practice Problems

Use the definitions of the different trigonometric functions and their values at the special angles to evaluate the following. Do not use a calculator.

1. $\sin(-\pi)$ __________

2. $\sin\left(\frac{3\pi}{4}\right)$ __________

3. $\sin\left(\frac{7\pi}{3}\right)$ __________

4. $\sin\left(-\frac{13\pi}{6}\right)$ __________

5. $\cos\left(-\frac{\pi}{4}\right)$ __________

6. $\cos(4\pi)$ __________

7. $\cos\left(\frac{4\pi}{3}\right)$ __________

8. $\tan\left(-\frac{5\pi}{3}\right)$ __________

9. $\tan\left(\frac{23\pi}{6}\right)$ __________

10. $\tan\left(-\frac{5\pi}{4}\right)$ __________

11. For which real x-values is the secant function not defined?

____

12. For which real x-values is the cosecant function not defined?

____

13. For which real x-values is the cotangent function not defined?

____

14. Complete the following table without using a calculator (instead, use your knowledge of basic trigonometric function values at the special angles).

θ	$0=0^\circ$	$\frac{\pi}{6}=30^\circ$	$\frac{\pi}{4}=45^\circ$	$\frac{\pi}{3}=60^\circ$	$\frac{\pi}{2}=90^\circ$
$\csc(\theta)$					
$\sec(\theta)$					
$\cot(\theta)$					

15. What is the range of the cosecant function?

____

16. What is the range of the secant function?

____

17. What is the range of the cotangent function?

____

Find the exact value of the following expressions without the use of a calculator.

18. $\csc\left(\frac{7\pi}{3}\right)$ ________

19. $\csc\left(-\frac{5\pi}{6}\right)$ ________

20. $\csc\left(\frac{3\pi}{2}\right)$ ________

21. $\sec(-\pi)$ ________

22. $\sec\left(\frac{11\pi}{6}\right)$ ________

23. $\sec\left(-\frac{4\pi}{3}\right)$ ________

24. $\sec\left(\frac{17\pi}{6}\right)$ ________

25. $\cot\left(-\frac{3\pi}{2}\right)$ ________

26. $\cot\left(\frac{11\pi}{4}\right)$ ________

27. $\cot\left(-\frac{5\pi}{6}\right)$ ________

28. $\cos(\pi)-3\sin(\pi/2)$ ________

29. $4\cos(\pi/4)-2\sin(\pi/6)$ ________

30. $\sin(\pi/3)-2\cos(5\pi/3)$ ________

31. $\tan(5\pi/6)-\sin(\pi/3)$ ________

32. $\frac{\tan(3\pi/4)-2\sin(-\pi)}{2}$ ________

33. $2\cos(-\pi)+2\sec(\pi/3)$ ________

34. $\csc(\pi/6)-3\sin(\pi/2)$ ________

35. $\csc(\pi/3)-2\cot(\pi/3)$ ________

36. $3\sec(\pi/6)+\cot(\pi/6)$ ________

37. $6\ \tan(\pi/3)-3\sec(\pi/6)$ ________

Appendix: Evaluating and Plotting Trigonometric Functions

Practice Problems

Use a graphing calculator to evaluate the following trigonometric functions. Round your answer to four decimal places.
(Note that you are to find trigonometric functions of angles in radians, so your calculator should be set to work in radians.)

1. $\sin\left(\frac{3\pi}{5}\right)$ ______ **2.** $\cos\left(-\frac{4\pi}{7}\right)$ ______

3. $\sec\left(-\frac{3\pi}{8}\right)$ ______ **4.** $\tan\left(\frac{2\pi}{9}\right)$ ______

5. $\csc\left(-\frac{9\pi}{5}\right)$ ______ **6.** $\cot\left(\frac{6\pi}{7}\right)$ ______

Use a graphing calculator to evaluate the following trigonometric functions. Round your answer to four decimal places.
(Note that you are to find trigonometric functions of angles in degrees, so your calculator should be set to work in degrees.)

7. $\sin\left(39^o\right)$ ______ **8.** $\cos\left(137^o\right)$ ______

9. $\sec\left(149^o\right)$ ______ **10.** $\tan\left(-314^o\right)$ ______

11. $\csc\left(208^o\right)$ ______ **12.** $\cot\left(-152^o\right)$ ______

Use a graphing calculator to graph each of the following trigonometric functions in the requested interval, and note the subintervals in which the function is increasing.

13. $\sin(x)$ in $[-\pi, 3\pi]$

14. $\cos(x)$ in $[-2\pi, 4\pi]$

15. $\tan(x)$ in $[-\pi, \pi]$

16. $\csc(x)$ in $[-2\pi, \pi]$

17. $\sec(x)$ in $[-2\pi, \pi]$

18. $\cot(x)$ in $\left[-\frac{3\pi}{2}, \frac{3\pi}{2}\right]$

Use a calculator to evaluate the following expressions. Round your answers to four decimal places.
(Be sure to set your calculator to work in radians.)

19. $\cot\left(\frac{5\pi}{8}\right) - 2\sec\left(\frac{7\pi}{5}\right)$ ______ **20.** $3\csc\left(\frac{3\pi}{8}\right) + 2\sec\left(\frac{6\pi}{5}\right)$ ______

21. $4\tan\left(\frac{7\pi}{9}\right) - 2\cot\left(\frac{9\pi}{5}\right)$ ______ **22.** $5\sin\left(\frac{3\pi}{7}\right) - 3\sec\left(\frac{10\pi}{7}\right)$ ______

Chapter 5 Section 3: Transformation of Trigonometric Functions

Practice Problems

Write the expression that shifts or alters the graph of each function as described.

1. $\sin(x)$ three units down. ______________________

2. $\tan(x)$ $\frac{\pi}{4}$ units to the left. ______________________

3. $\cos(x)$ $\frac{\pi}{4}$ units to the right and $\frac{5}{2}$ units up. ______________________

4. $\csc(x)$ $\frac{\pi}{2}$ units to the left and $\frac{1}{2}$ unit up. ______________________

5. $\cot(x)$ $\frac{3}{2}$ units to the right and $\frac{1}{10}$ unit down. ______________________

6. $\sec(x)$ a units to the left and 2.04 units down. ______________________

7. Stretches the graph of the function $\sin(x)$ vertically four times. ______________________

8. Reflects the graph of the function $\csc(x)$ around the y-axis. ______________________

9. Shrinks the graph of the function $\tan(x)$ vertically three times. ______________________

10. Reflects the graph of the function $\sec(x)$ around the x-axis. ______________________

11. Stretches the graph of the function $\cos(x)$ 2.5 times vertically and reflects it around the x-axis. ______________________

12. Shrinks the graph of the function $\cot(x)$ vertically in half and reflects it around the y-axis. ______________________

13. Stretches the graph of the function $\sin(x)$ horizontally three times. ______________________

14. Shrinks the graph of the function $\sec(x)$ horizontally by $1/4$. ______________________

15. Stretches the graph of the function $\cos(x)$ horizontally 4.5 times and reflects it around the x-axis. ______________________

16. Shrinks the graph of the function $\csc(x)$ horizontally in half and reflects it around the y-axis. ______________________

Study the functions below and determine the period of each, without graphing.

17. $\sin(3x)$ __________

18. $\cos\left(\frac{x}{4}\right)$ __________

19. $\tan(2x)$ __________

20. $\cot\left(\frac{1}{5}x\right)$ __________

21. $\frac{1}{2}\sec(8x)$ __________

22. $3\csc\left(\frac{x}{2}\right)$ __________

Find the expression of a function of the form $f(x) = A\ \sin(bx)$ *that gives each graph below.*

23.

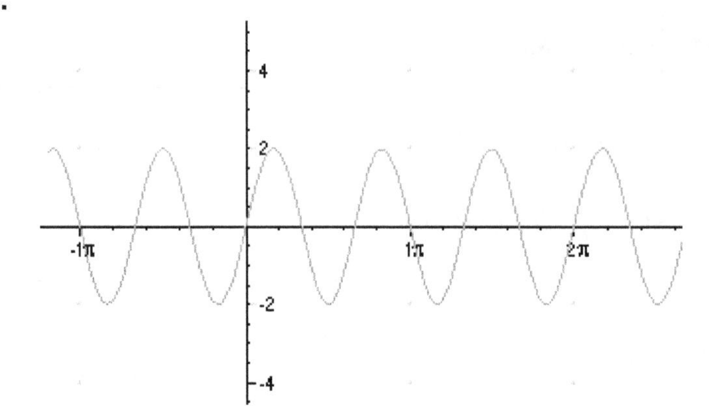

____

24.

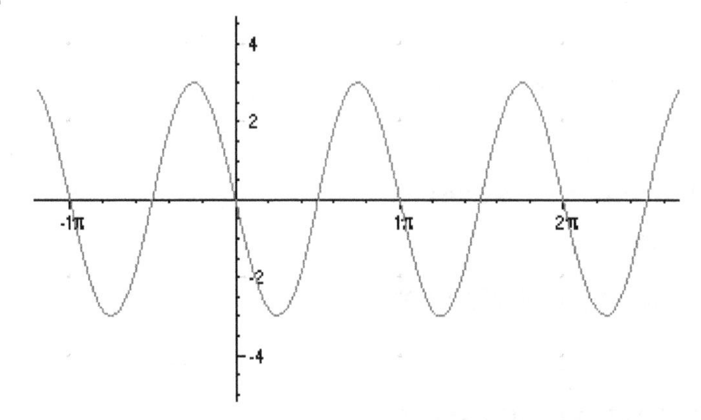

____

25.

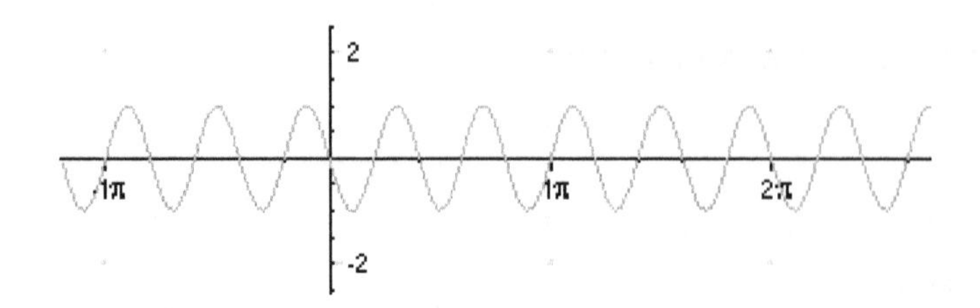

____

26.

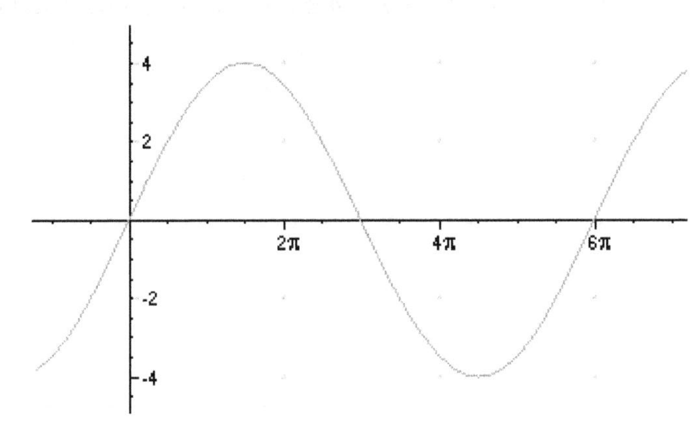

____

Find the expression of a function of the form $f(x) = A\ \cos(bx)$ *that gives each graph below.*

27.

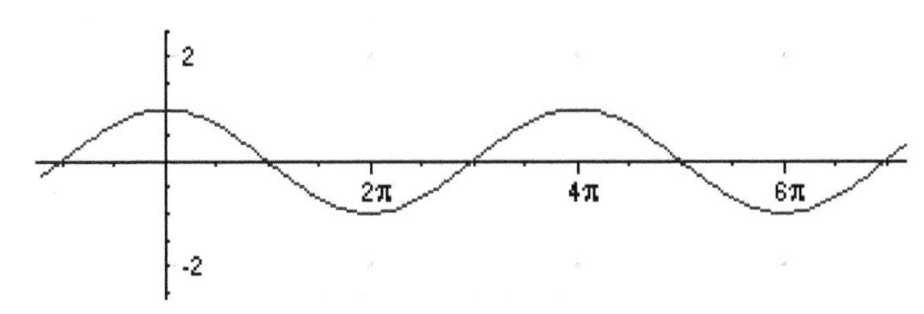

____

28.

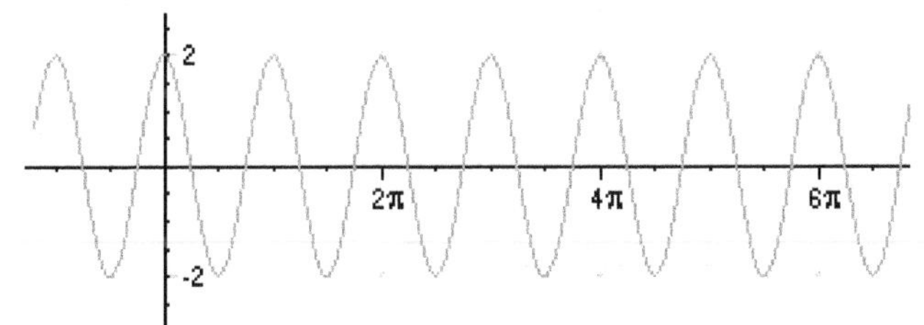

____

29.

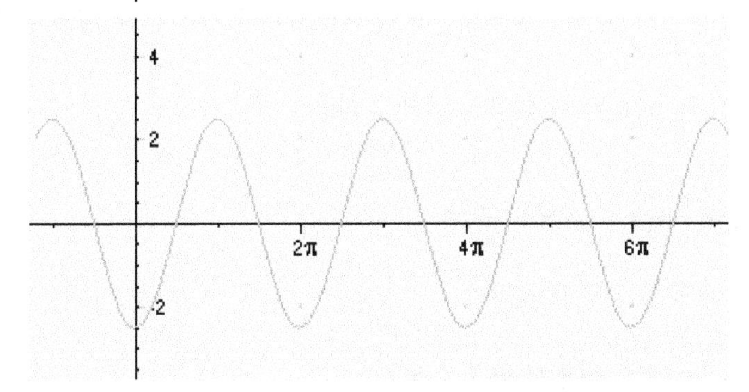

____

30.

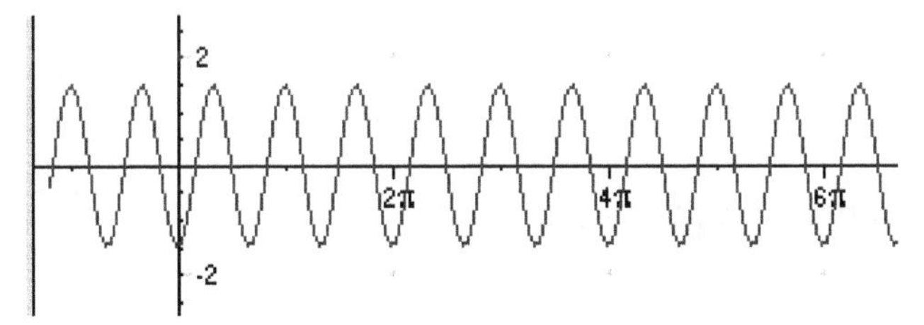

____

Find the amplitude, period, and phase shift of the harmonic motion described by each of the following functions.

31. $f(x)=4\cos(3x-\pi)$

---

32. $f(x)=-2.5\cos(2x+2\pi)$

---

33. $f(x)=-\frac{2}{5}\cos\left(4x-\frac{\pi}{2}\right)$

---

34. $f(x)=-\sin\left(3x+\frac{\pi}{2}\right)$

---

35. $f(x)=\frac{1}{9}\sin\left(\frac{x-\pi}{5}\right)$

Find an expression for the harmonic motion described in each case below, using a cosine function of the form:

$$f(x) = A\cos(bx + c)$$

36. Amplitude = 3, period = π, and phase shift = $\pi/2$

---

37. Amplitude = $\frac{1}{2}$, period = $\pi/3$, and phase shift = $-\pi$

---

38. Amplitude = 5, period = 4π , and phase shift = -2π

---

39. Amplitude = $\frac{3}{2}$, period = $\pi/2$, and phase shift = π

____

40. Amplitude = $\frac{1}{4}$, period = $3\pi/2$, and phase shift = $-\pi/3$

____

Find the period of each of the following combinations of trigonometric functions. Visually check your answers with a graphing calculator.

41. $2\cos(x)+\sin(4x)$ ________

42. $3\sin(2x)-\cos(3x)$ ________

43. $-2\cos(6x)+\sin\left(\frac{x}{2}\right)$ ________

44. $\sin(5x)-2\cos\left(\frac{x}{3}\right)$ ________

45. $-\frac{1}{2}\sin(6x)+\frac{1}{3}\cos(3x)$ ________

Use a graphing calculator to graph the following functions in the requested interval.

46. $f(x)=\sin(x)\cdot\sin(10x)$ in $[-2\pi,2\pi]$

47. $f(x)=\sin(2x)\cdot\cos(24x)$ in $[-2\pi,2\pi]$

48. $f(x)=x^2\cdot\cos(12x)$ in $[-2\pi,2\pi]$

49. $f(x)=\dfrac{\sin(6x)}{x}$ in $[-2\pi,2\pi]$

50. $f(x)=\left(x^3-2x\right)\cdot\sin(8x)$ in $[-2\pi,2\pi]$

Chapter 5 Section 4: Algebra and Identities of Trigonometric Functions

Practice Problems

Find each product and write your answer using compact power notation for trigonometric functions.

1. $\cos(\theta)\left[1-\cos(\theta)\right]\left[\cos(\theta)+2\right]$ ____________________

2. $\left[3+\sin(\alpha)\right]\cdot\left[2\sin(\alpha)-4\right]$ ____________________

3. $\left[2\tan(x)-3\right]\cdot\left[2\tan(x)+3\right]$ ____________________

4. $\left[5-2\sin(\beta)\right]^2$ ____________________

5. $\left[3\cos(x)-\sin(y)\right]\cdot\left[2\sin(y)+4\cos(x)\right]$ ____________________

Factor the following expressions.

6. $18\sin(\alpha)\cos(\beta)-3\sin^2(\alpha)$ ____________________

7. $4-\cos^2(\theta)$ ____________________

8. $6\sin^2(\beta)+\sin(\beta)-2$ ____________________

9. $64\tan^2(\theta)-1$ ____________________

10. $\cos^4(\alpha)+2\cos^2(\alpha)-3$ ____________________

Use the negative angle identities to find the exact value of each of the following.

11. $\cos(-\theta)$ knowing that $\cos(\theta)=\frac{2}{5}$ ____________________

12. $\sin(-\beta)$ knowing that $\sin(\theta)=\frac{\sqrt{5}}{8}$ ____________________

13. $\sec(x)$ knowing that $\sec(-x)=-\frac{\sqrt{2}}{5}$ ____________________

14. $\cot(\theta)$ knowing that $\cot(-\theta)=-\frac{\sqrt{21}}{4}$ ____________________

Use a Pythagorean identity to find the exact value of each of the following.

15. $\cos(x)$ knowing that $\sin(x)=\frac{2}{7}$ and that $0<x<\frac{\pi}{2}$ (angle x is in the first quadrant)

____

16. $\sin(y)$ knowing that $\cos(y)=-\frac{\sqrt{3}}{6}$ and that $\frac{\pi}{2}<y<\pi$ (angle y is in the second quadrant)

____

17. $\tan(\theta)$ knowing that $\sec(\theta)=\frac{8}{3}$ and that $\frac{\pi}{2}<\theta<\pi$ (angle θ is in the second quadrant)

____

18. $\csc(x)$ knowing that $\cot(x)=-\frac{\sqrt{5}}{10}$ and that $\frac{3\pi}{2}<x<2\pi$ (angle x is in the fourth quadrant)

____

19. $\tan(\theta)$ knowing that $\cos(\theta)=-\frac{\sqrt{6}}{3}$ and that $\pi<\theta<\frac{3\pi}{2}$ (angle θ is in the third quadrant)

____

Use algebra and the identities of trigonometric functions to simplify each of the following expressions. Assume that denominators are NOT zero.

20. $\frac{\cot(\theta)}{15}\cdot 12\sec(\theta)$ ______________

21. $3\tan(x)-2\sec(x)$ ______________

22. $\frac{\cos^2(y)+2\cos(y)-3}{\cos(y)-1}$ ______________

23. $\frac{1-3\sin(\theta)-\cos^2(\theta)}{\sin(\theta)}$ ______________

24. $\csc^2(x)\left[\frac{\sin(x)}{\sec(x)}-\tan(x)\right]$ ______________

25. $\sec(x)\cdot\tan\left(\frac{\pi}{2}-x\right)$ ______________

26. $\cot\left(\frac{\pi}{2}-y\right)\cdot\csc(y)\cdot\sin\left(\frac{\pi}{2}-y\right)$ ______________

Use the addition or the subtraction identities to find the exact value of each of the following.

27. $\sin\left(75^o\right)$ ____________

28. $\cos\left(105^o\right)$ ____________

29. $\tan\left(\frac{\pi}{12}\right)$ ____________

30. $\sin\left(\frac{13\pi}{12}\right)$ ____________

31. $\cos\left(\frac{11\pi}{12}\right)$ ____________

Use the addition and/or subtraction identities to simplify each of the following in terms of trigonometric functions of the individual angles.

32. $\sin\left(x+\pi\right)$ ____________

33. $\cos\left(x+\frac{3\pi}{2}\right)$ ____________

34. $\tan\left(x-\frac{\pi}{4}\right)$ ____________

35. $\csc\left(x-\frac{\pi}{3}\right)$ ____________

36. $\sec(x-3\pi)$ ____________

Find the quadrant in which the angle $\theta/2$ resides in each case given.

37. θ in first quadrant ____________

38. θ in second quadrant ____________

39. θ in third quadrant ____________

40. θ in fourth quadrant ____________

Use the half-angle identities to find the exact value of each of the following.

41. $\sin\left(\frac{\pi}{8}\right)$ ____________

42. $\cos\left(\frac{3\pi}{8}\right)$ ____________

43. $\tan\left(\frac{5\pi}{8}\right)$ ____________

44. $\csc\left(\frac{11\pi}{12}\right)$ ____________

Find the trigonometric value of each given function.

45. $\cos\left(\frac{x}{2}\right)$ knowing that x is an angle in the first quadrant and $\cos\left(x\right)=\frac{1}{5}$

____

46. $\sin\left(\frac{x}{2}\right)$ knowing that x is an angle in the third quadrant and $\cos(x) = -\frac{3}{5}$

____

47. $\tan\left(\frac{x}{2}\right)$ knowing that x is an angle in the fourth quadrant and $\cos(x) = \frac{3}{8}$

____

48. $\cos(2x)$ knowing that x is an angle in the first quadrant and $\cos(x) = \frac{1}{6}$

____

49. $\sin(2x)$ knowing that x is an angle in the first quadrant and $\cos(x) = \frac{1}{4}$

____

Appendix: More Elaborate Trigonometric Identities

Practice Problems

Write each of the following products as a sum of sines or cosines.

1. $\cos(4\theta)\cdot\sin(3\theta)$

____

2. $\sin(5\theta)\cdot\sin(2\theta)$

____

3. $\cos(8\theta)\cdot\cos(4\theta)$

____

4. $\sin(\theta)\cdot\cos(5\theta)$

____

5. $\sin\left(\frac{3\theta}{2}\right)\cdot\cos\left(\frac{\theta}{2}\right)$

____

6. $\sin\left(\frac{4\theta}{3}\right)\cdot\sin\left(\frac{5\theta}{3}\right)$

____

Write each of the following expressions as a product of sines and/or cosines.

7. $\sin(2\alpha)+\sin(6\alpha)$

____

8. $\sin(3\alpha)-\sin(7\alpha)$

____

9. $\cos(4\alpha)-\cos(2\alpha)$

---

10. $\cos(\alpha)+\cos(5\alpha)$

---

11. $\sin\left(\frac{3\alpha}{2}\right)-\sin\left(\frac{5\alpha}{2}\right)$

---

12. $\cos\left(\frac{7\alpha}{2}\right)-\cos\left(\frac{\alpha}{2}\right)$

Chapter 5 Section 5: Proving Trigonometric Identities

Practice Problems

Use a graphing calculator to decide if the following equations may represent identities.

1. $\cos(2x)-\sin^2(x)=\cos^2(x)$ ________________

2. $\sin(\pi-x)=\cos\left(x-\frac{\pi}{2}\right)$ ________________

3. $\frac{1+\cos(x)}{\sin(x)}=\sec(x)\cdot\tan(x)$ ________________

4. $[\sin(x)+\cos(x)]^2=1$ ________________

5. $\sec(x)\cdot\csc(x)=\tan(x)+\cot(x)$ ________________

6. $1+\csc^2(x)=\cot^2(x)$ ________________

7. $1-\sin(x)=\frac{\cos^2(x)}{1+\sin(x)}$ ________________

8. $\sec(x)-\tan(x)=1+\cot(x)$ ________________

Prove each of the following identities. Write your proofs on a separate sheet of paper.

9. $\cos(6\theta)=\cos^2(3\theta)-\sin^2(3\theta)$

10. $\sin(2x)=\frac{2\tan(x)}{1+\tan^2(x)}$

11. $1-\sin^2(x)=\cot^2(x)\cdot\sin^2(x)$

12. $\sin^2(x)+\sin^2(x)\cdot\tan^2(x)=\tan^2(x)$

13. $\tan^2(x)[\cot(x)+1]^2=[\tan(x)+1]^2$

14. $\frac{1+\sin(x)-\sin^2(x)}{\cos(x)}=\cos(x)+\tan(x)$

15. $[1-\tan(x)]\cdot[1+\cot(x)]=[2-\sec^2(x)]\cot(x)$

16. $\csc(x)=\frac{\sec(x)+\csc(x)}{1+\tan(x)}$

17. $\dfrac{1-\cot(x)}{\tan(x)-1}=\cot(x)$

18. $\dfrac{\sec(x)-\csc(x)}{1-\tan(x)}=-\csc(x)$

19. $\dfrac{1-\sin(x)}{\cos(x)}=\dfrac{\cos(x)}{1+\sin(x)}$

20. $\dfrac{\sin(x)}{1+\cos(x)}=\dfrac{1-\cos(x)}{\sin(x)}$

21. $\dfrac{\cos(x)}{1-\sin(x)}=\tan(x)+\sec(x)$

22. $\tan(x)+\cot(x)=\sec(x)\cdot\csc(x)$

23. $\left[1+\cot(x)\right]\sec(x)=\sec(x)+\csc(x)$

24. $\dfrac{\csc(x)-\sec(x)}{\tan(x)-1}=-\csc(x)$

25. $\dfrac{\cos(x-y)}{\sin(y)\cos(x)}=\tan(x)+\cot(y)$

Chapter 5 Section 6: Solving Trigonometric Equations

Practice Problems

Without a calculator, evaluate the following inverse functions for the specified values. Give your answer in radians and in degrees. (You may use a calculator to check your results).

1. $\arcsin(1)$ ____________

2. $\arccos(-1)$ ____________

3. $\arctan(0)$ ____________

4. $\arcsin\left(-\frac{\sqrt{3}}{2}\right)$ ____________

5. $\arccos\left(\frac{1}{2}\right)$ ____________

6. $\arctan\left(-\frac{\sqrt{3}}{3}\right)$ ____________

7. $\arcsin\left(-\frac{1}{2}\right)$ ____________

8. $\arccos\left(-\frac{\sqrt{2}}{2}\right)$ ____________

9. $\arctan\left(\sqrt{3}\right)$ ____________

10. $\arccos(0)$ ____________

Use a calculator to evaluate the following inverse functions at the specified values. Give your answer in radians and in degrees, rounded to two decimal places.

11. $\arcsin\left(\frac{2}{7}\right)$ ____________

12. $\arccos\left(\frac{3}{4}\right)$ ____________

13. $\arctan\left(\frac{8}{3}\right)$ ____________

14. $\arcsin\left(-\frac{\sqrt{5}}{9}\right)$ ____________

15. $\arccos\left(-\frac{\sqrt{3}}{7}\right)$ ____________

16. $\arctan(-5)$ ____________

17. $\arcsin(-0.3447)$ ____________

18. $\arccos(0.0154)$ ____________

19. $\arctan(0.1975)$ ____________

20. $\arctan(-7.809)$ ____________

Without the use of a calculator, find the exact solution to each of the following compositions of trigonometric functions, in radians.

21. $\arccos(\sin(0))$ ________

22. $\arcsin\left(\tan\left(\frac{3\pi}{4}\right)\right)$ ________

23. $\arcsin\left(\cos\left(\frac{5\pi}{4}\right)\right)$ ________

24. $\arctan\left(\tan\left(\frac{7\pi}{6}\right)\right)$ ________

25. $\arccos\left(\sin\left(\frac{4\pi}{3}\right)\right)$ ________

Use a calculator to evaluate the following compositions of trigonometric functions. Round your answers to three decimal places.

26. $\sin\left(\arcsin\left(-\frac{2}{5}\right)\right)$ ________

27. $\cos\left(\arcsin\left(\frac{1}{6}\right)\right)$ ________

28. $\tan\left(\arccos\left(-\frac{\sqrt{2}}{3}\right)\right)$ ________

29. $\sin\left(\arccos\left(-\frac{\sqrt{5}}{6}\right)\right)$ ________

30. $\cos\left(\arctan\left(-\frac{10}{3}\right)\right)$ ________

31. $\arccos\left(\cos\left(\frac{3\pi}{7}\right)\right)$ ________

32. $\arcsin\left(\cos\left(\frac{3\pi}{5}\right)\right)$ ________

33. $\arctan\left(\sin\left(\frac{7\pi}{3}\right)\right)$ ________

34. $\arccos\left(\tan\left(32^o\right)\right)$ ________

35. $\arcsin\left(\sin\left(158^o\right)\right)$ ________

36. $\arctan\left(\sin\left(328^o\right)\right)$ ________

37. $\arccos\left(\tan\left(195^o\right)\right)$ ________

Find all solutions to each of the following equations. Give exact answers.

38. $\cos(x)=0$ ________________

39. $2\sin(x)-1=0$ ________________

40. $\tan(4x)=-\sqrt{3}$ ________________

41. $\csc(3x)+1=0$ ________________

42. $\sec^2(x)=1$ ________________

43. $8\sin^3(x)+1=0$ ________________

44. $3\tan^2(x)-1=0$ ________________

45. $4\sin^2(x)-3=0$ ________________

46. $\sin^2(x)=\sin(x)$ ________________

47. $2\cos^2(x)+\cos(x)-1=0$ ________________

Use a calculator to help you find the solutions to the following equations within the given interval. Round your answers to four decimal places.

48. $4\tan(x) = -15$ for $-\infty < x < \infty$ ______________________

49. $3\cos(x) = 2$ for $0 \le x \le 2\pi$ ______________________

50. $5\sin\left(x - \frac{\pi}{2}\right) = 1$ for $-\pi \le x \le \pi$ ______________________

51. $\cos\left(\frac{x}{3}\right) = -0.6$ for $0 \le x \le 2\pi$ ______________________

52. $\frac{x}{2} + 3 = 4\sin(2x)$ for $-\pi \le x \le \pi$ ______________________

53. $\tan(x) = 4\sin(x) + 2$ for $-\frac{\pi}{2} \le x \le \frac{\pi}{2}$ ______________________

54. $\cot(x) = 2 + \sin\left(\frac{x}{2}\right)$ for $-\frac{\pi}{2} \le x \le \frac{\pi}{2}$ ______________________

55. $3\sin(2x) + 4\cos(2x) = 4$ for $-\infty < x < \infty$ ______________________

56. $3\sin^2(x) = \cos\left(\frac{x}{2}\right)$ for $-\frac{\pi}{2} \le x \le \frac{\pi}{2}$ ______________________

57. $\sin^2(x) = \cos^2(x) - 2$ for $-\infty < x < \infty$ ______________________

CONTENT on Demand

PRECALCULUS 2ND EDITION

Practice Problem Worksheets

CHAPTER 6: APPLICATIONS OF TRIGONOMETRY

Chapter 6 Section 1: Right Triangle Trigonometry

Practice Problems

The terminal side of an angle in standard position passes through the given point on the plane. Find the three basic trigonometric functions for the angle in each case below.

1. $\left(3, \sqrt{7}\right)$ ________________________________

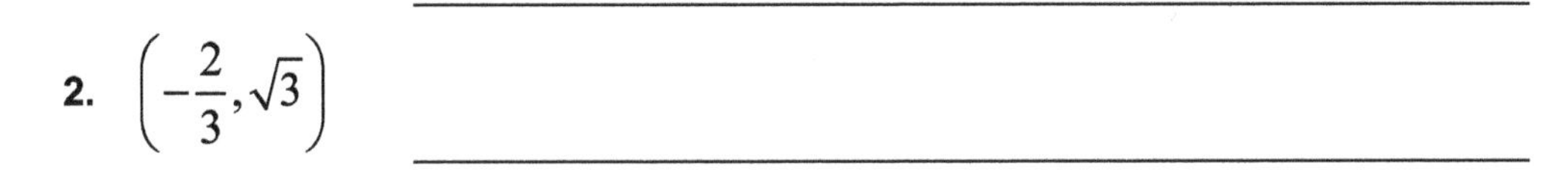

2. $\left(-\frac{2}{3}, \sqrt{3}\right)$ ________________________________

3. $\left(\sqrt{5}, -\sqrt{2}\right)$ ________________________________

4. $\left(-\frac{1}{10}, \frac{2}{5}\right)$ ________________________________

5. $\left(-\frac{\sqrt{6}}{5}, -\sqrt{3}\right)$ ________________________________

The terminal side of an angle in standard position passes through the given point on the plane. Find the angle measure in degrees, rounded to one decimal place.

6. $\left(\frac{3}{4}, \sqrt{3}\right)$ ________________

7. $\left(-\frac{4}{3}, \sqrt{2}\right)$ ________________

8. $\left(-2\sqrt{3}, -5\right)$ ________________

9. $\left(\frac{5}{2}, -\frac{\sqrt{5}}{4}\right)$ ________________

10. $\left(-\sqrt{5}, 0.1\right)$ ________________

Find all six trigonometric functions of angle θ for each right triangle shown.

11.

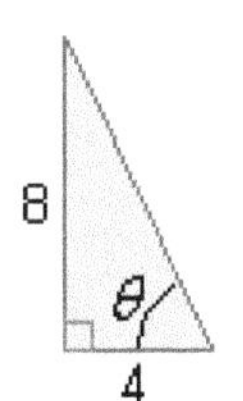

____

12.

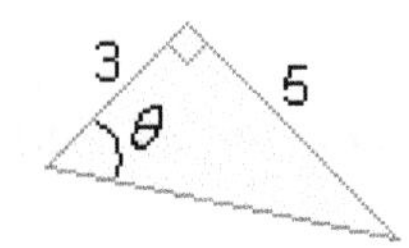

13.

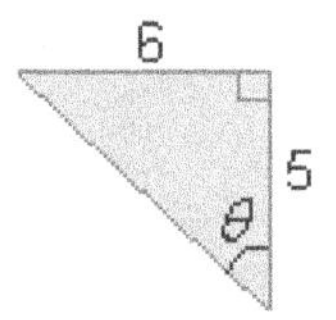

14.

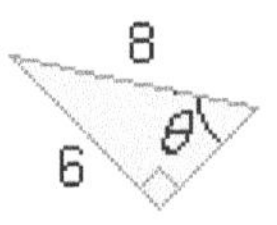

15.

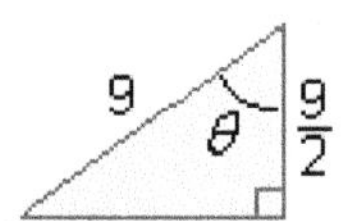

Find the value of the indicated angle θ for each right triangle depicted below. Give your answer in degrees, rounded to one decimal place.

16.

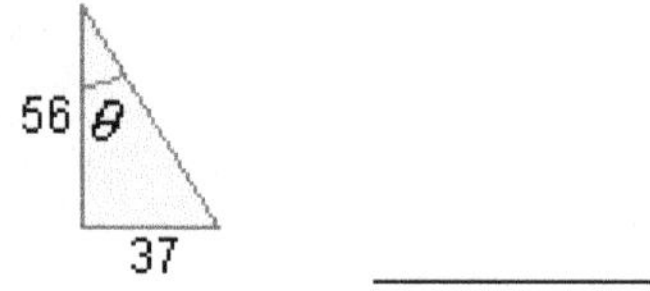

17.

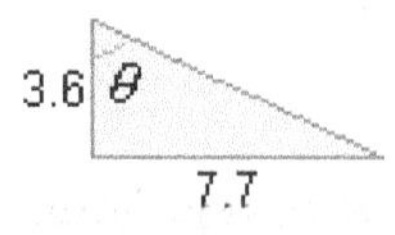

18.

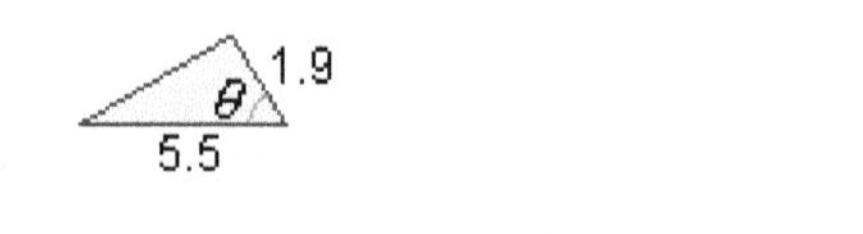

19.

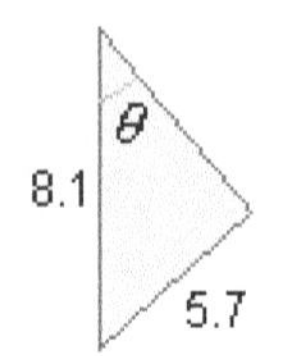

20.

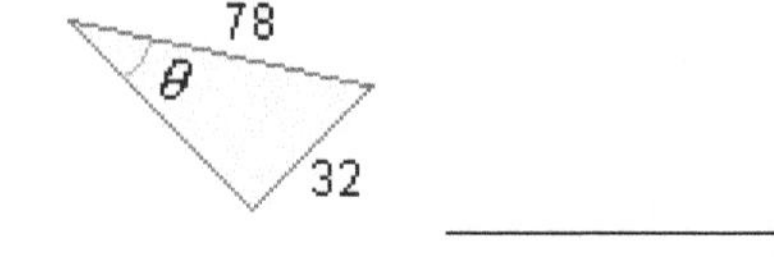

21.

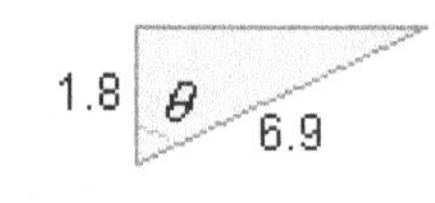

Use the appropriate trigonometric ratios to find the requested unknown in the following right triangles. Round your answers to four decimal places, if necessary.

22.

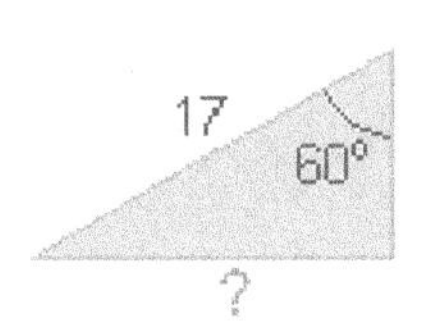

____

23.

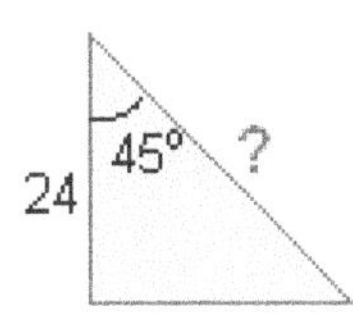

____

24.

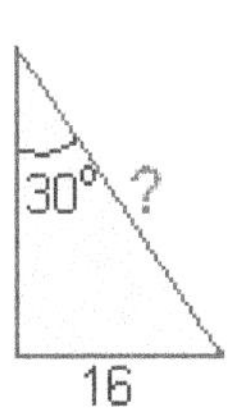

____

25.

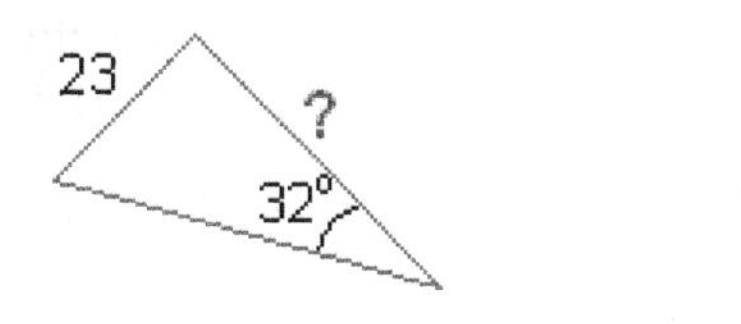

____

26.

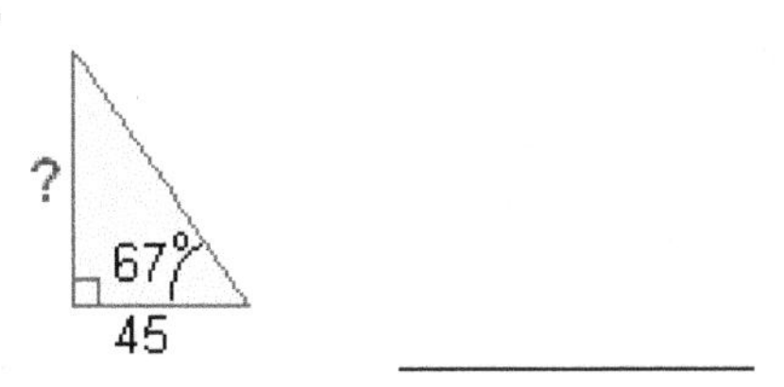

____

27. A person has to stand a twelve-foot ladder against the wall of a building to do some maintenance work. In order to avoid damaging some ornamental plants, he has to keep the base of the ladder at a distance of at least 5.6 feet from the wall. What is the maximum inclination angle (θ) the ladder can have, and what is the distance from the ground to the top of the ladder in this situation?

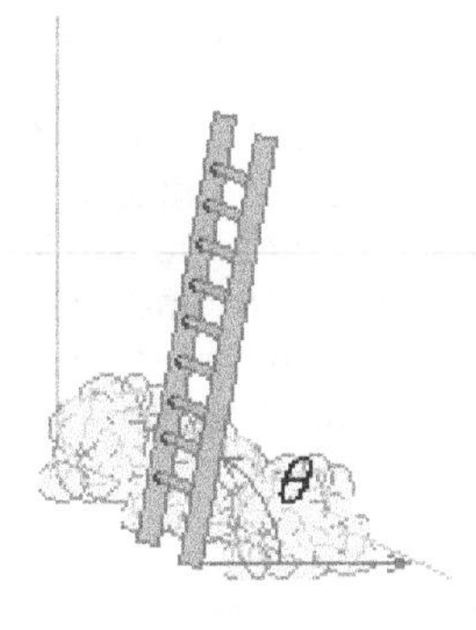

____

28. A camera is placed at an angle of elevation of 27.5° to photograph a hot air balloon suspended above the ground at a horizontal distance of 4.8 miles from the camera. How far above the ground is the balloon?

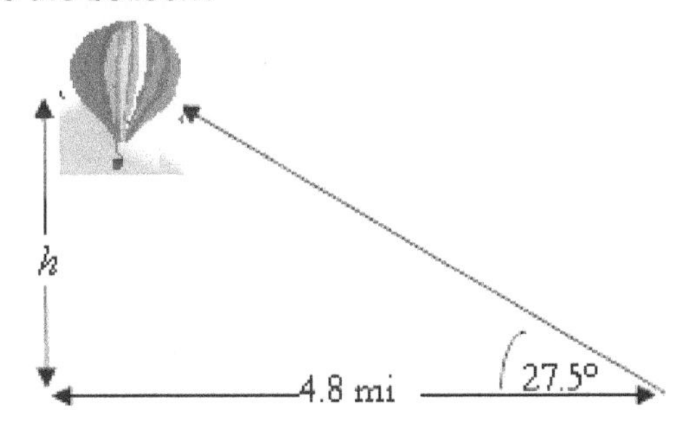

____

29. The beams for a roof need to connect two parallel walls separated by a distance of 10 feet, as indicated in the figure. If one of the walls is 12 feet high and the other one is 8 feet high, what is the minimum length of the beams to be used? What is the roof's angle of elevation?

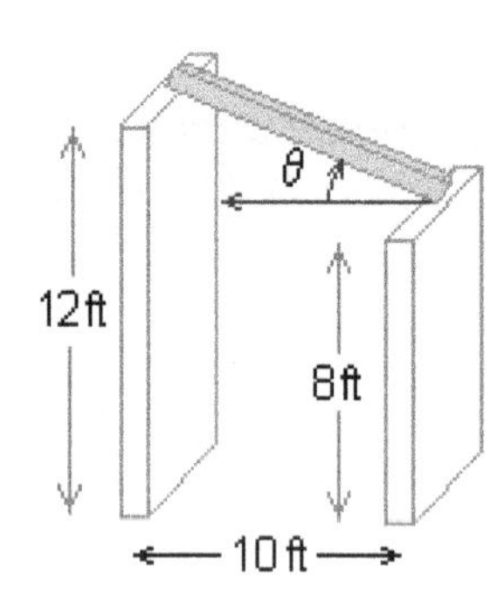

---

30. A person standing on the bank of a river sees a big statue on the edge of the opposite bank. The angle of depression from the person's eyes to the bottom of the statue's platform is 18°, while the angle of elevation to the top of the statue is 26°. Considering that the person's eyes are 5.6 feet from the ground, how tall is the statue, including the platform (h in the image at the right)? How far away from the statue is the person standing (d)?

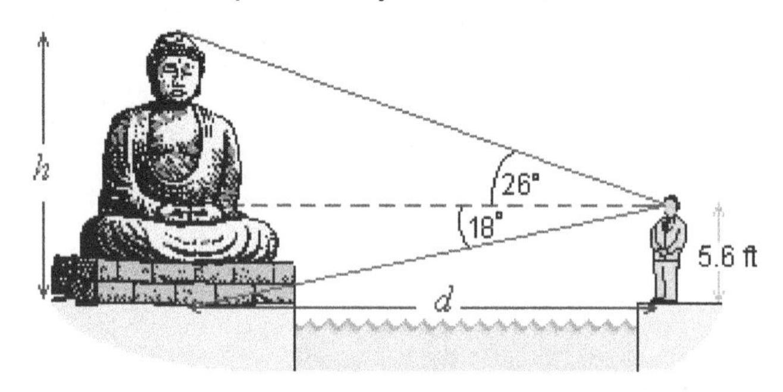

---

31. A survey plane flying at an altitude of 31,000 feet is measuring the diameter of a newly deforested circular area in the tropics. It simultaneously records the two limit angles for the borders of the circular area at 75° and 132°, measured with respect to the direction of flight, as shown in the image. Find the diameter of the deforested region.

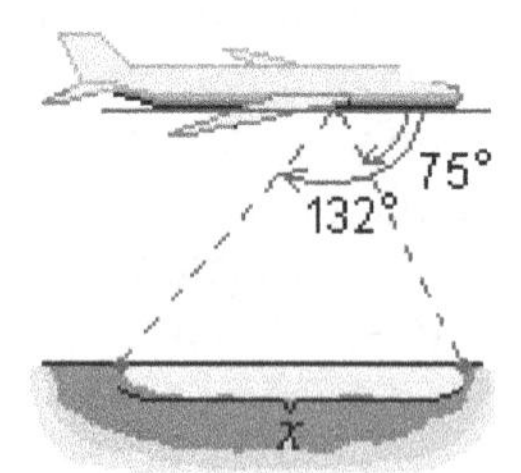

Chapter 6 Section 2: The Law of Sines and the Law of Cosines

Practice Problems

Solve each of the following oblique triangles.

1. $\hat{A}=61^o$ $\quad a=?$
   $\hat{B}=?$ $\quad b=?$ →
   $\hat{C}=14^o$ $\quad c=3.7$

   B
   c a
   A b C
   known values

   ____

2. $\hat{A}=?$ $\quad a=?$
   $\hat{B}=22^o$ $\quad b=45$ →
   $\hat{C}=114^o$ $\quad c=?$

   B
   c a
   A b C
   known values

   ____

3. $\hat{A}=24^o$ $\quad a=?$
   $\hat{B}=?$ $\quad b=52$ →
   $\hat{C}=57^o$ $\quad c=?$

   B
   c a
   A b C
   known values

   ____

4. $\hat{A}=18^o$ $\quad a=?$
   $\hat{B}=121^o$ $\quad b=?$ →
   $\hat{C}=?$ $\quad c=8.5$

   B
   c a
   A b C
   known values

   ____

5. $\hat{A}=74^o$ $\quad a=2.2$
   $\hat{B}=?$ $\quad b=?$ →
   $\hat{C}=?$ $\quad c=4.9$

   B
   c a
   A b C
   known values

   ____

6. $\hat{A}=?$ $\quad a=74$
   $\hat{B}=103^o$ $\quad b=51$ →
   $\hat{C}=?$ $\quad c=?$

   B
   c a
   A b C
   known values

   ____

7. $\hat{A}=?$ $\quad a=?$
   $\hat{B}=?$ $\quad b=38$ →
   $\hat{C}=97^o$ $\quad c=54$

   B
   c a
   A b C
   known values

   ____

8.

$\hat{A} = 56^o$ $a = 6.9$
$\hat{B} = ?$ $b = ?$ →
$\hat{C} = ?$ $c = 7.8$

B
c a
A b C
known values

____

9.

$\hat{A} = ?$ $a = 63$
$\hat{B} = 49^o$ $b = 77$ →
$\hat{C} = ?$ $c = ?$

B
c a
A b C
known values

____

10.

$\hat{A} = ?$ $a = ?$
$\hat{B} = ?$ $b = 25$ →
$\hat{C} = 34^o$ $c = 18$

B
c a
A b C
known values

____

11.

$\hat{A} = 79^o$ $a = 6.7$
$\hat{B} = ?$ $b = ?$ →
$\hat{C} = ?$ $c = 5.8$

B
c a
A b C
known values

____

12.

$\hat{A} = 39^o$ $a = ?$
$\hat{B} = ?$ $b = 9.2$ →
$\hat{C} = ?$ $c = 6.8$

B
c a
A b C
known values

____

13.

$\hat{A} = ?$ $a = 34$
$\hat{B} = ?$ $b = 27$ →
$\hat{C} = 118^o$ $c = ?$

B
c a
A b C
known values

____

14.

$\hat{A} = ?$ $a = 10.6$
$\hat{B} = ?$ $b = 9.4$ →
$\hat{C} = ?$ $c = 6.8$

B
c a
A b C
known values

____

15.

$\hat{A} = ?$ $a = 66$
$\hat{B} = ?$ $b = 28$ →
$\hat{C} = ?$ $c = 43$

B
c a
A b C
known values

____

Calculate the area of each triangle defined below.

16. $\hat{A} = 37^o \quad b = 5.5 \quad c = 4.6$ ____________

17. $\hat{A} = 87^\circ \quad a = 41 \quad c = 28$ ____________

18. $\hat{A} = 138^o \quad \hat{C} = 16^o \quad b = 8.4$ ____________

19. $\hat{A} = 142^\circ \quad c = 9.6 \quad a = 16$ ____________

20. $\hat{B} = 75^o \quad \hat{C} = 32^o \quad b = 9.2$ ____________

21. $a = 7 \quad b = 4 \quad c = 5$ ____________

22. $a = 13 \quad b = 16 \quad c = 24$ ____________

23. $a = 3.9 \quad b = 7.1 \quad c = 8.4$ ____________

24. A trucker driving on a straight road sees a tall grain silo in the distance, at an angle of 32° from the direction he is traveling. Five miles farther down the road, the trucker now sees the silo at an angle of 85° from the direction of travel. How far away is the truck from the silo at that instant?

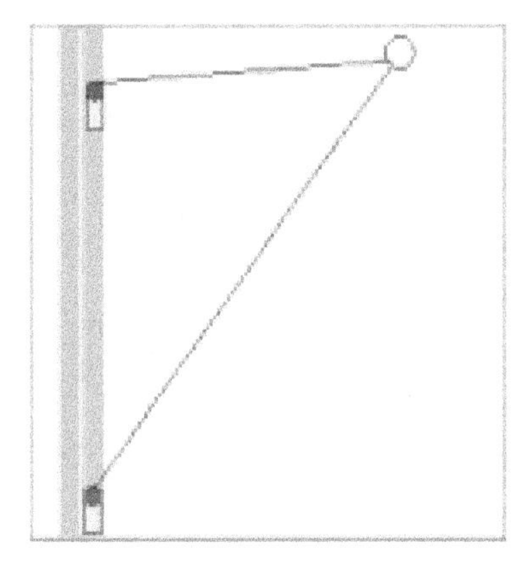

____

25. A boat sails north from the harbor for 4.5 miles an then makes a 16° deviation east from the original direction of travel. After traveling 6.5 miles in this new direction, how far is the boat from the harbor?

____

26. A cable 52 feet long runs from an anchor on the ground to the top of an antenna subtending an angle of elevation of 54°. From the same anchor a second cable is run to a point 18 feet down from the antenna's top, subtending an angle of elevation of 38°. Find the length of the second cable. Round your answer to two decimal places.

____

Practice Problems

Eight points on the plane are shown in the image below. Use the polar grid to find a pair of polar coordinates to represent each point, with $0 \le \theta < 2\pi$ and positive "r."

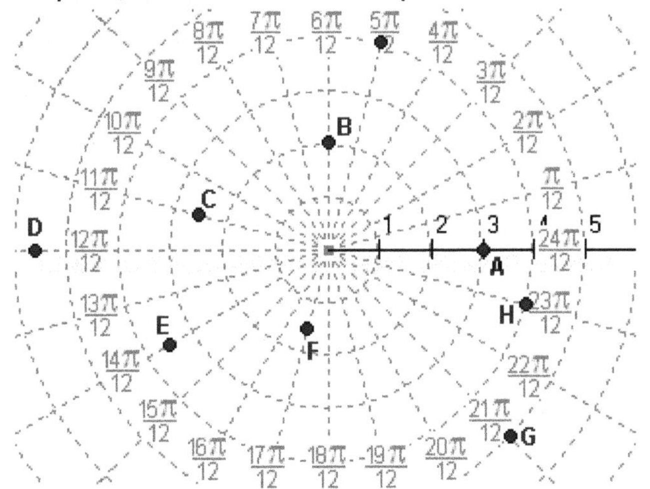

1. **A** ________________

2. **B** ________________

3. **C** ________________

4. **D** ________________

5. **E** ________________

6. **F** ________________

7. **G** ________________

8. **H** ________________

Plot each of the following points in the polar coordinate system.

9. $\left(\frac{9}{2}, \frac{5\pi}{4}\right)$

10. $\left(-\frac{5}{2},\frac{\pi}{3}\right)$

11. $\left(-4,\frac{7\pi}{6}\right)$

12. $\left(-\frac{3}{2},-\frac{7\pi}{12}\right)$

13. $\left(\frac{7}{2},-\frac{13\pi}{6}\right)$

Find an equivalent pair of polar coordinates for the given points and with the following conditions: positive "r" and $0 \le \theta < 2\pi$.

14. $\left(-\frac{3}{2},\frac{\pi}{3}\right)$ ________

15. $\left(-\frac{1}{2},\frac{8\pi}{3}\right)$ ________

16. $\left(\frac{5}{3},-\frac{8\pi}{3}\right)$ ________

17. $\left(-\frac{7}{4},\frac{9\pi}{4}\right)$ ________

18. $\left(-\frac{9}{5},-\frac{11\pi}{3}\right)$ ________

Convert the following points given in polar coordinates into rectangular coordinates.

19. $\left(2,\frac{5\pi}{6}\right)$ ________

20. $\left(1,\frac{11\pi}{2}\right)$ ________

21. $\left(\frac{1}{2},-\frac{3\pi}{4}\right)$ ________

22. $\left(-2,-\frac{2\pi}{3}\right)$ ________

23. $\left(-\frac{1}{3},-\frac{11\pi}{6}\right)$ ________

Use a calculator to convert the following points given in polar coordinates into rectangular coordinates. Round your answer to three decimal places.

24. $\left(4,\frac{\pi}{8}\right)$ ________

25. $\left(\frac{1}{3},\frac{2\pi}{5}\right)$ ________

26. $\left(-2,\frac{4\pi}{7}\right)$ ________

27. $\left(-\frac{3}{2},-\frac{\pi}{5}\right)$ ________

28. $\left(-0.4,-\frac{7\pi}{8}\right)$ ________

Write the following points given in rectangular coordinates in polar coordinates, with positive r and with θ values within the interval $[0, 2\pi)$.

29. $(-4, 0)$ ____________

30. $\left(0, -\frac{5}{2}\right)$ ____________

31. $\left(\frac{3\sqrt{3}}{8}, \frac{3}{8}\right)$ ____________

32. $\left(-4\sqrt{2}, 4\sqrt{2}\right)$ ____________

33. $\left(3, -3\sqrt{3}\right)$ ____________

Convert each of the following equations given in rectangular coordinates into an equation in polar coordinates.

34. $x = -5$ ____________

35. $y = 3x - 4$ ____________

36. $x^2 + y^2 = 16$ ____________

37. $3x^2 - 5y^2 = 2$ ____________

38. $\frac{x^2 + y^2}{2} = \frac{3x}{y}$ ____________

Convert each of the following equations given in polar coordinates into an equation in rectangular coordinates.

39. $r \cdot \sin(\theta) = -2$ ____________

40. $r^2 = 3$ ____________

41. $r = 5$ ____________

42. $r = 3 + \sin(\theta)$ ____________

43. $r = \frac{6}{1 + \cos(\theta)}$ ____________

44. $r = 3 \cdot \tan(\theta)$ ____________

45. $3 \cdot \csc(\theta) = 2\sqrt{3}$ ____________

Use a graphing calculator to graph the following polar equations, with θ varying within $[0, 2\pi]$.

46. $r = 6\sin(\theta)$

47. $r = 1 + 2\sin(\theta)$

48. $r = 4 - 3\cos(\theta)$

49. $r = 3.5\sin(2\theta)$

50. $r = \sin(4\theta) + \cos(\theta)$

51. $r = \sin(3\theta) + \cos(2\theta)$

Practice Problems

Find the modulus (absolute value) of each of the following complex numbers.

1. $-1+\sqrt{3}\ i$ ________

2. $-2-\sqrt{5}\ i$ ________

3. $4-i$ ________

4. $-\frac{3}{2}+\frac{\sqrt{5}}{2}\ i$ ________

5. $-\frac{1}{4}-\frac{\sqrt{7}}{2}\ i$ ________

Write the following complex numbers in polar form.

6. $-5+5\ i$ ________________

7. $4\sqrt{3}+4\ i$ ________________

8. $\frac{2}{3}-\frac{2}{3}\ i$ ________________

9. $-5-5\sqrt{3}\ i$ ________________

10. $\frac{3\sqrt{2}}{2}-\frac{\sqrt{6}}{2}\ i$ ________________

Write the following complex numbers in polar form. Give the exact value for the modulus and round the trigonometric part to the nearest degree, using an angle within $\left[0,360^\circ\right)$.

11. $2+6\ i$ ________________________________

12. $-\frac{4}{3}-\ i$ ________________________________

13. $\sqrt{11}-5\ i$ ________________________________

14. $-\sqrt{2}+\sqrt{7}\ i$ ________________________________

15. $-\frac{\sqrt{10}}{2}-\sqrt{3}\ i$ ________________________________

Find the product of the following complex numbers, in polar form.

16. $2\left[\cos\left(\frac{3\pi}{2}\right)+i\sin\left(\frac{3\pi}{2}\right)\right]$ and $3\left[\cos\left(\frac{\pi}{2}\right)+i\sin\left(\frac{\pi}{2}\right)\right]$ ________________

17. $5\left[\cos\left(\frac{3\pi}{5}\right)+i\sin\left(\frac{3\pi}{5}\right)\right]$ and $\frac{3}{10}\left[\cos\left(\frac{6\pi}{5}\right)+i\sin\left(\frac{6\pi}{5}\right)\right]$ ________________

18. $\sqrt{6}\left[\cos\left(\frac{\pi}{8}\right)+i\sin\left(\frac{\pi}{8}\right)\right]$ and $\sqrt{15}\left[\cos\left(\frac{7\pi}{8}\right)+i\sin\left(\frac{7\pi}{8}\right)\right]$ ________________

19. $\frac{2}{7}\left[\cos\left(15^o\right)+i\sin\left(15^o\right)\right]$ and $\frac{14}{3}\left[\cos\left(108^o\right)+i\sin\left(108^o\right)\right]$ ________________

20. $\frac{3\sqrt{5}}{2}\left[\cos\left(39^o\right)+i\sin\left(39^o\right)\right]$ and $\frac{\sqrt{10}}{3}\left[\cos\left(18^o\right)+i\sin\left(18^o\right)\right]$ ________________

Use De Moivre's Theorem to find the following powers of complex numbers. Work with angles in degrees within the interval $\left[0^\circ,360^\circ\right)$ and round them to one decimal place, when needed.

21. $\left\{\sqrt{2}\left[\cos\left(72^o\right)+i\ \sin\left(72^o\right)\right]\right\}^5$ ________________

22. $\left\{\sqrt{5}\left[\cos\left(123^o\right)+i\ \sin\left(123^o\right)\right]\right\}^8$ ________________

23. $(3+3\ i)^5$ ________________

24. $\left(3-\sqrt{3}\ i\right)^4$ ________________

25. $\left(-2+\sqrt{5}\ i\right)^6$ ________________

26. $\left(-\sqrt{13}-\sqrt{3}\ i\right)^5$ ________________

Find the requested roots of the complex number given. Use θ in degrees and round it to one decimal place, when needed.

27. Complex cube roots of $27\left[\cos\left(120^o\right)+i\ \sin\left(120^o\right)\right]$

____

28. Complex fifth roots of $32\left[\cos\left(30^o\right)+i\ \sin\left(30^o\right)\right]$

____

29. Complex cube roots of -1

---

30. Complex fourth roots of $81\,i$

---

31. Complex cube roots of $(-8-2i)$

---

32. Complex fourth roots of $(-3+4i)$

Practice Problems

Find the magnitude of the vector $\overrightarrow{\mathbf{AB}}$, given the following rectangular coordinates for points A and B.

1. $A=(2,5)$ and $B=(6,8)$ ______

2. $A=(-1,8)$ and $B=(-6,2)$ ______

3. $A=(1,-4)$ and $B=(-3,-6)$ ______

4. $A=\left(-\frac{1}{2},-3\right)$ and $B=\left(\frac{5}{2},-1\right)$ ______

5. $A=\left(\sqrt{2},-\frac{5}{3}\right)$ and $B=\left(-3\sqrt{2},-\frac{2}{3}\right)$ ______

6. Sketch three vectors equivalent to the one shown, using initial and final points different from the ones for $\overrightarrow{\mathbf{AB}}$.

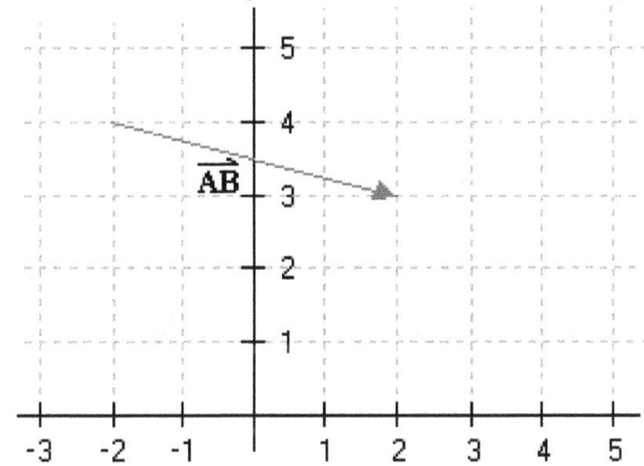

Find the component form of the vector $\overrightarrow{\mathbf{AB}}$ given the (x,y) coordinates of the points below.

7. $A=(-1,5)$ and $B=(3,6)$ ______

8. $A=(-3,4)$ and $B=(-3,-10)$ ______

9. $A=(4,-9)$ and $B=(-3,-7)$ ______

10. $A=\left(-\frac{3}{2},-2\right)$ and $B=\left(\frac{5}{2},-5\right)$ ______

11. $A=\left(-\sqrt{3},-\frac{5}{8}\right)$ and $B=\left(-3\sqrt{3},-\frac{7}{8}\right)$ ______

Find the magnitude of each of the following vectors given in component form.

12. $\langle 2,4\rangle$ ______

13. $\langle -4,9\rangle$ ______

14. $\left\langle -3,\frac{1}{2}\right\rangle$ ______

15. $\left\langle -\frac{5}{3},\frac{7}{3}\right\rangle$ ______

16. $\langle 2\sqrt{5},-\sqrt{15}\rangle$ ______

Find the component form and magnitude of each of the vectors shown in the following diagram.

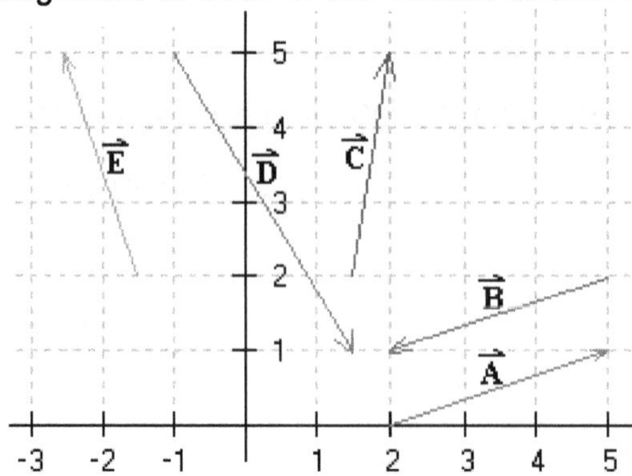

17. Vector A ____________________

18. Vector B ____________________

19. Vector C ____________________

20. Vector D ____________________

21. Vector E ____________________

Given the component form of vectors $\vec{u}$ and $\vec{v}$ given below, find the components of the vector resulting from each indicated operation.

$$\vec{u}=\left\langle -3.5,\frac{9}{4}\right\rangle \text{ and } \vec{v}=\left\langle 6,-\frac{3}{5}\right\rangle$$

22. $\vec{u}+\vec{v}$ ____________________

23. $\vec{v}-\vec{u}$ ____________________

24. $-\frac{5}{6}\vec{v}$ ____________________

25. $-2\vec{u}+3\vec{v}$ ____________________

26. $-\frac{1}{2}\vec{u}-\frac{1}{4}\vec{v}$ ____________________

27. Find the component form of the airplane's velocity vector in the given rectangular coordinate system, knowing that the magnitude of the velocity is 550 miles per hour and that the angle θ of elevation is 18°. Round your answer to the nearest whole number of miles per hour.

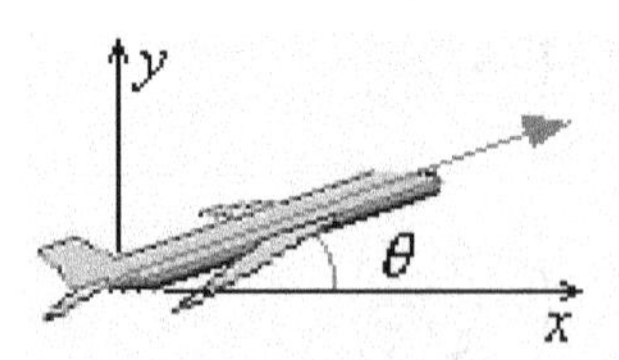

____

28. Find the component form of the package's weight force vector (w) in the given rectangular coordinate system, knowing that its weight is 15 pounds and that the angle of inclination of the plane (θ) is 21°. Give your answer in pounds, rounded to two decimal places.

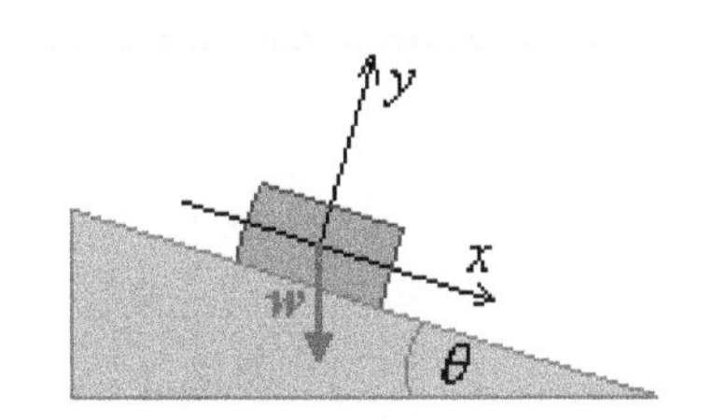

____

Find the vector resulting from the addition of the vectors shown. Give your answer in component form using the given coordinate system.

29.

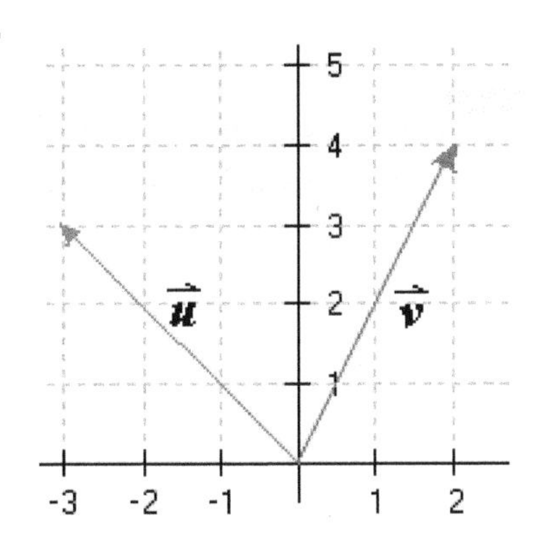

____

30.

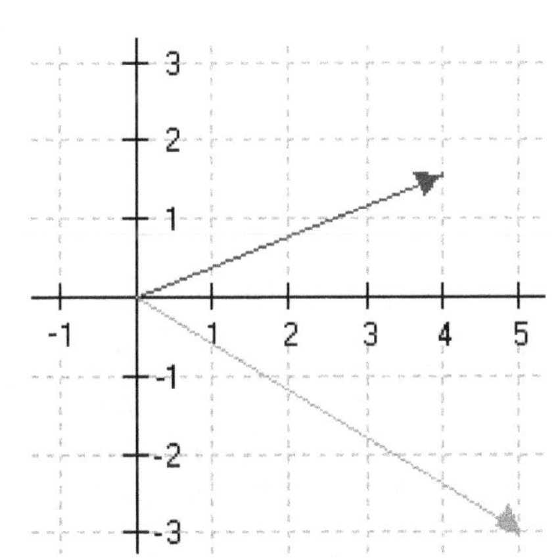

____

31.

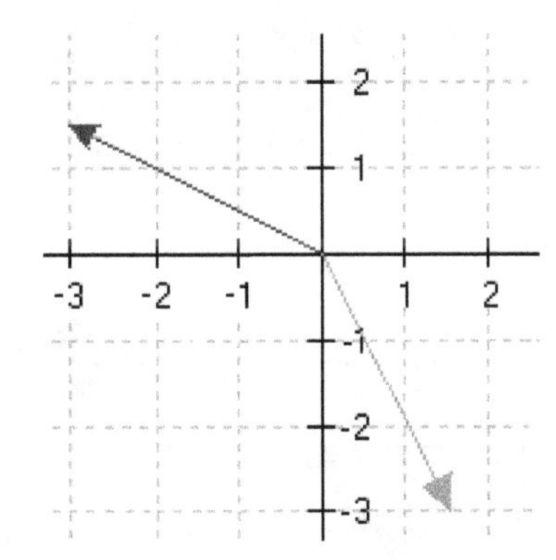

____

32.

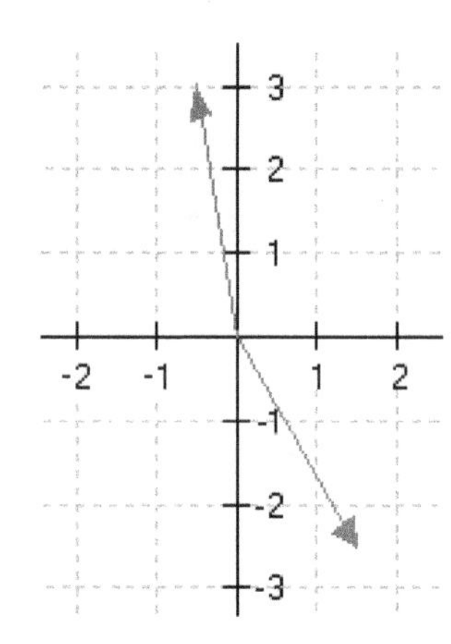

____

33. A bird is flying directly east with a speed in still air of 25 miles per hour. Find the bird's effective speed when the atmospheric conditions change as in each case given below. Note: The wind is always blowing *from* the given direction and not toward it.

a. East wind at 10 miles per hour.

____

b. West wind at 8 miles per hour.

____

c. North wind at 12 miles per hour.

____

d. Southwest wind at 15 miles per hour.

____

34. A person needs to get directly across a river by boat. The boat's speed is 20 feet per second. There is a strong river current of 14 ft per second. Towards which direction should the person head the boat to keep a straight path across the river? What is the magnitude of the resulting velocity vector (see image at right)?

____

35. Two tractors are pulling a heavy cart to a silo. One of the tractors exerts a force of 4500 pounds and the other, smaller tractor, a force of 3900 pounds. The pulling ropes define an angle θ of 46°. Find the magnitude of the resulting force (in pounds) applied to the cart. Round your answer to two decimal places.

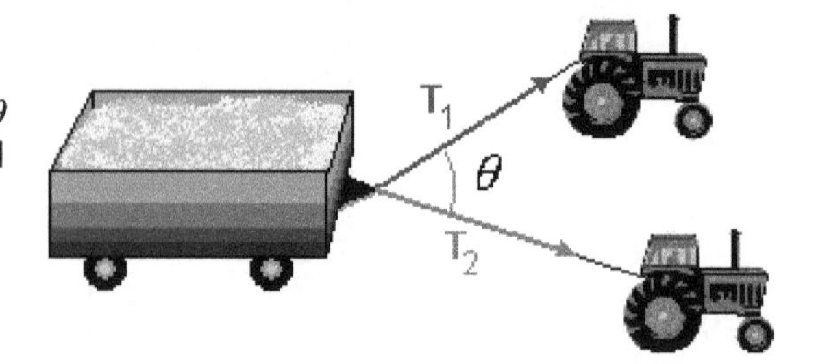

____

36. A 150-pound object is suspended by two ropes as indicated in the figure. Find the magnitude of the tension exerted on each rope (T_1 and T_2 in the figure) if their respective angles of elevation are $\alpha = 42^\circ$ and $\beta = 31^\circ$. Round your answers to one-tenth of a pound.

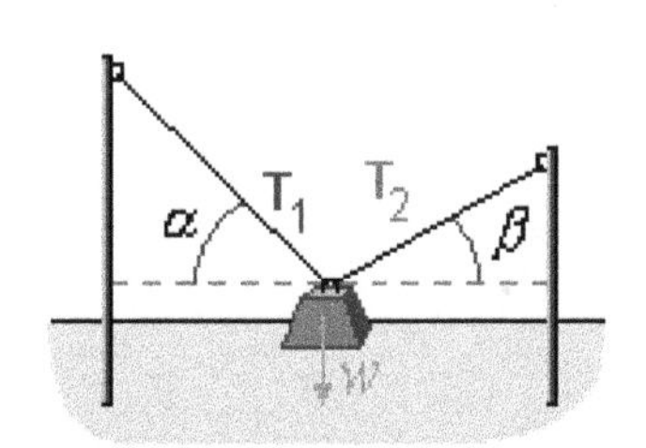

CONTENT on Demand

Precalculus 2nd Edition

Practice Problem Worksheets

Chapter 7: Analytic Geometry

Practice Problems

Graph the following parametric equations by creating a table and plotting the points.

1. $\begin{cases} x(t) = 2t - 1 \\ y(t) = t^2 + 3t \end{cases}$ for $-10 \le t \le 10$

---

2. $\begin{cases} x(t) = 2t^3 + 4t \\ y(t) = 5 - 4t \end{cases}$ for $-10 \le t \le 10$

---

3. $\begin{cases} x(t) = -t + 1 \\ y(t) = \sqrt{3t^2 + 3} \end{cases}$ for $-10 \le t \le 10$

---

4. $\begin{cases} x(t) = 3t - 2 \\ y(t) = \sqrt{4 - t^2} \end{cases}$ for $-2 \le t \le 2$

---

5. $\begin{cases} x(t) = 3t - 5 \\ y(t) = \dfrac{2}{t} \end{cases}$ for $-10 \le t < 0$ and $0 < t \le 10$

---

6. $\begin{cases} x(t) = \dfrac{5}{\sqrt{t}} \\ y(t) = t^3 - 4t \end{cases}$ for $0 < t \le 10$

---

7. $\begin{cases} x(t) = \sqrt{16-t^2} \\ y(t) = t+3 \end{cases}$ for $-4 \le t \le 4$

---

8. $\begin{cases} x(t) = \sqrt{2t} \\ y(t) = e^{t-1} \end{cases}$ for $0 \le t \le 10$

---

9. $\begin{cases} x(t) = 2^t \\ y(t) = 3t^2 - t - 5 \end{cases}$ for $-2 \le t \le 4$

---

10. $\begin{cases} x(t) = \left(\dfrac{1}{2}\right)^{t-2} \\ y(t) = t^2 - 5t \end{cases}$ for $-2 \le t \le 7$

Obtain an equivalent equation in rectangular coordinates for each set of parametric equations.

11. $\begin{cases} x(t) = 5t - 2 \\ y(t) = 2 - t \end{cases}$ ____________________

12. $\begin{cases} x(t) = -3t - 4 \\ y(t) = 5t + 1 \end{cases}$ ____________________

13. $\begin{cases} x(t) = 2t + 3 \\ y(t) = 3t^2 - t \end{cases}$ ____________________

14. $\begin{cases} x(t) = 5t^2 + 2 \\ y(t) = -3t + 1 \end{cases}$ ____________________

15. $\begin{cases} x(t) = 2^{t-1} \\ y(t) = 3t - 4 \end{cases}$ ____________________

16. $\begin{cases} x(t) = 2t + 1 \\ y(t) = \dfrac{3}{t-5} \end{cases}$ ____________________

17. $\begin{cases} x(t) = \dfrac{5}{t} \\ y(t) = t^2 - 4t + 2 \end{cases}$ ____________________

18. $\begin{cases} x(t) = \sqrt{t^2 + 2t} \\ y(t) = -3t + 1 \end{cases}$ ____________________

19. $\begin{cases} x(t) = t^2 + t - 7 \\ y(t) = \sqrt{t} \end{cases}$ for $t > 0$ ____________________

20. $\begin{cases} x(t) = \dfrac{3}{2-t} \\ y(t) = e^{-t+1} \end{cases}$ ____________________

Find a set of parametric equations that produce the same graph as the given function or line segment below.

21. $f(x) = 3x - 2 \quad -10 \le x \le 10$ ____________________

22. $f(x) = \dfrac{-x+3}{2} \quad -10 \le x \le 10$ ____________________

23. The line segment joining the points (0,3) and (5,8) ____________________

24. The line segment joining the points (−1,4) and (2,−5) ____________________

25. $f(x) = 3x^2 - 5x + 1 \quad -10 \le x \le 10$ ____________________

26. $f(x) = \sqrt{x-5} \quad 5 \le x$ ____________________

27. $f(x) = \sqrt{9 - x^2} \quad -3 \le x \le 3$ ____________________

28. $f(x) = \dfrac{3x+7}{x+5} \quad x < -5$ ____________________

29. $f(x) = \dfrac{4x+1}{2x} \quad x > 0$ ____________________

30. $f(x) = e^{-\frac{(x-1)^2}{3}} \quad -5 \le x \le 5$ ____________________

Chapter 7 Section 2: Conics–Parabolas

Practice Problems

Find the focus, directrix, and focal diameter, and sketch the graph of each of the following parabolas.

1. $y^2 = 8x$

2. $y^2 = 2x$

3. $x^2 = y$

4. $\frac{x^2}{6} = -y$

5. $-y^2 = 5x$

6. $3x^2 = 4y$

7. $8x = -3y^2$

8. $24y - 2x^2 = 0$

9. $4y^2 - 14x = 0$

10. $5x^2 + 28y = 0$

For each problem below, find the standard equation for a parabola with vertex at the origin of coordinates (0,0) and with the given geometrical properties.

11. Focus at $(0,3)$

12. Focus at $\left(0,-\dfrac{1}{4}\right)$

13. Focus at $(-5,0)$ ________________________

14. Focus at $\left(\frac{1}{2},0\right)$ ________________________

15. Focus at $\left(0,\frac{9}{2}\right)$ ________________________

16. Directrix $x=7$ ________________________

17. Directrix $y=2$ ________________________

18. Directrix $y=-5$ ________________________

19. Directrix $x=\frac{1}{4}$ ________________________

20. Directrix $x=-\frac{1}{6}$ ________________________

21. Vertical axis of symmetry and the point $\left(1,\frac{1}{6}\right)$ is on the parabola's graph.

____

22. Horizontal axis of symmetry and the point $(-8,4)$ is on the parabola's graph.

____

23. Horizontal axis of symmetry and the point $\left(\frac{5}{6},-\sqrt{5}\right)$ is on the parabola's graph.

____

24. Vertical axis of symmetry and the point $\left(-3,\frac{9}{7}\right)$ is on the parabola's graph.

____

Match each parabola equation with its graph (note that there are choices on the next page). Write the letter of the correct answer in the box next to each equation.

25. $x^2 = -10y$ ☐

A.

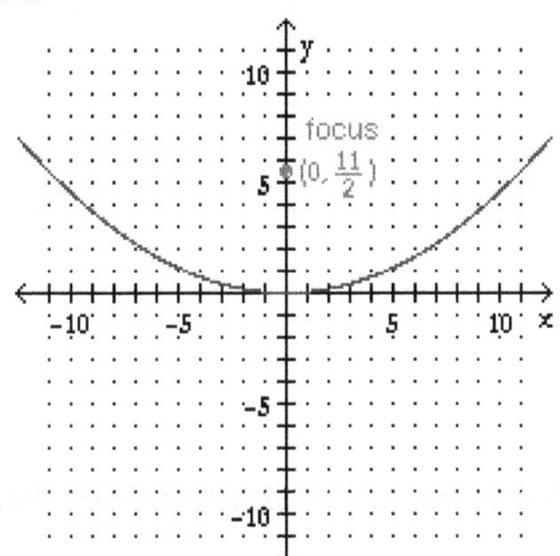

26. $y^2 = 16x$ ☐

B.

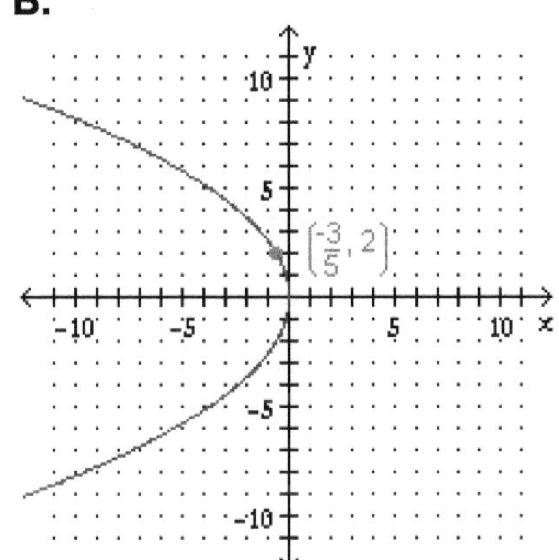

27. $x^2 - 10y = 0$ ☐

C.

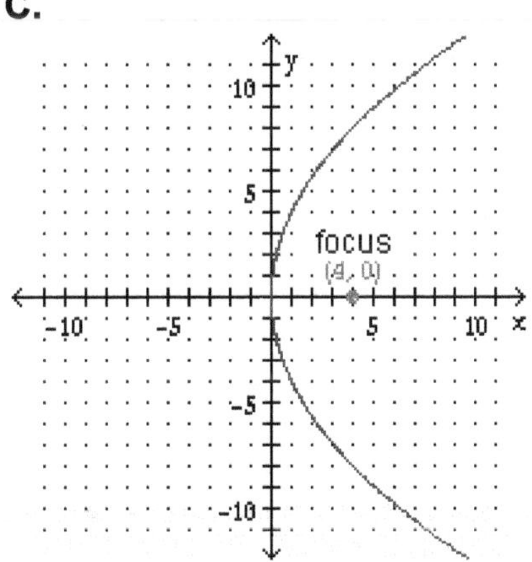

28. $2y^2 + 28x = 0$ ☐

D.

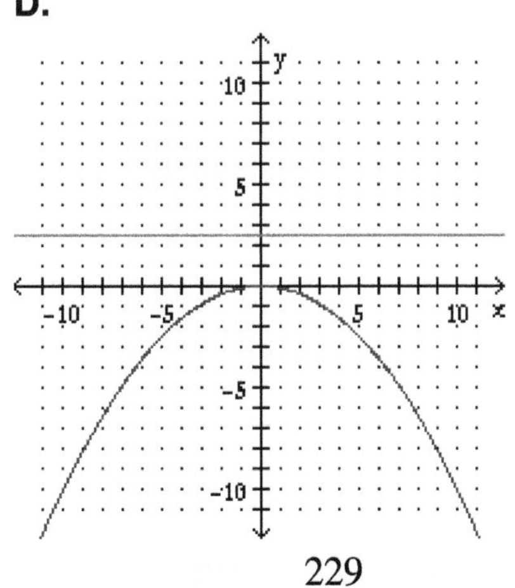

29. $x^2 - 22y = 0$ ☐

E.

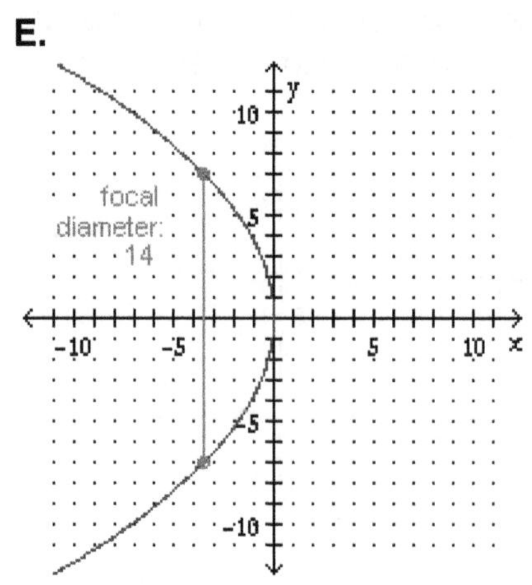

30. $3y^2 + 20x = 0$ ☐

F.

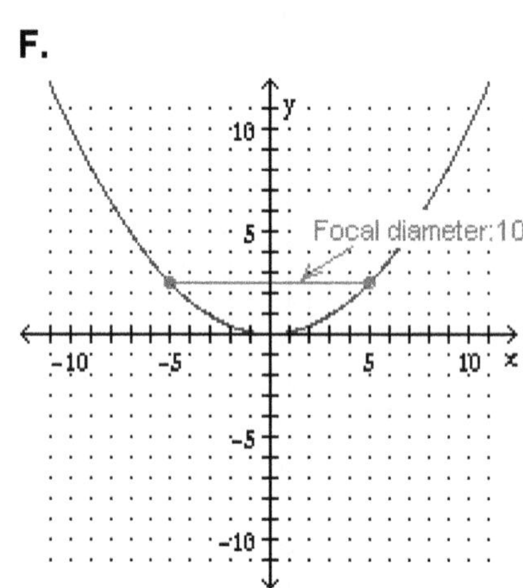

Practice Problems

Rewrite the equation of each ellipse given below, in standard form. State the parameters a and b, state whether the ellipse's principal axis is horizontal or vertical, and note the coordinates of the foci.

1. $16x^2 + y^2 = 16$

2. $x^2 + 25y^2 = 25$

3. $9x^2 + 16y^2 = 144$

4. $x^2 + 36y^2 = 36$

5. $25x^2 + 4y^2 = 100$

6. $2x^2 + y^2 = 8$

7. $5x^2 + 9y^2 = 45$

8. $2x^2 + y^2 = 6$

9. $2x^2 + 3y^2 = 12$

10. $7x^2 + 5y^2 = 70$

11. $4x^2 + y^2 = 1$

12. $25x^2 + 5y^2 = 1$

13. $3x^2 + 10y^2 = 1$

____

14. $4x^2 + 24y^2 = 4$

____

Plot the graph of each ellipse defined below. Start by finding parameters a and b and use them to find all four vertices. Then plot the vertices and sketch the graph. Also state the location of the foci.

15. $\frac{x^2}{9} + \frac{y^2}{4} = 1$

16. $\frac{x^2}{9} + \frac{y^2}{25} = 1$

17. $\frac{x^2}{100} + \frac{y^2}{25} = 1$

18. $x^2 + 4y^2 = 1$

19. $16x^2 + y^2 = 1$

20. $25x^2 + 36y^2 = 1$

21. $9x^2 + 2y^2 = 18$

22. $7x^2 + 16y^2 = 112$

23. $2x^2 + y^2 = 10$

Find the eccentricity of each ellipse defined below.

24. $\frac{x^2}{36} + \frac{y^2}{4} = 1$ ____________________

25. $\frac{x^2}{25} + \frac{y^2}{100} = 1$ ____________________

26. $4x^2 + y^2 = 1$ ____________________

27. $x^2 + 25y^2 = 1$ ____________________

28. $3x^2 + 24y^2 = 8$ ____________________

Find the standard equation of each ellipse below, centered at the origin, with the given eccentricity and foci.

29. $e = \frac{\sqrt{3}}{2}$ and foci: $\left(0, -3\sqrt{3}\right)$ and $\left(0, 3\sqrt{3}\right)$ ____________________

30. $e = \frac{3}{5}$ and foci: $(-3, 0)$ and $(3, 0)$ ____________________

31. $e = \frac{\sqrt{3}}{3}$ and foci: $\left(\sqrt{2}, 0\right)$ and $\left(\sqrt{2}, 0\right)$ ____________________

32. $e = \frac{\sqrt{2}}{2}$ and foci: $\left(0, -2\sqrt{2}\right)$ and $\left(0, 2\sqrt{2}\right)$ ____________________

33. $e = \frac{1}{2}$ and foci: $(-2, 0)$ and $(2, 0)$ ____________________

Chapter 7 Section 4: Conics–Hyperbolas

Practice Problems

Rewrite the equation of each hyperbola given below, in standard form. State the following: (a) the parameters a and b; (b) whether the hyperbola's transverse axis is horizontal or vertical; (c) the coordinates of the foci; (d) the coordinates of the vertices; and (e) the expression of the asymptotes.

1. $\frac{x^2}{9} - \frac{y^2}{25} = 1$

____

2. $\frac{y^2}{16} - \frac{x^2}{25} = 1$

____

3. $9x^2 - 4y^2 = 36$

____

4. $25y^2 - 4x^2 = 100$

____

5. $x^2 - 4y^2 = 16$

____

6. $25y^2 - x^2 = 25$

____

7. $x^2 - 2y^2 = 4$

____

8. $y^2 - 4x^2 = 2$

____

9. $5x^2 - 6y^2 = 30$

____

10. $y^2 - 2x^2 = 1$

____

Graph each hyperbola below by writing the equation in standard form to find the main parameters and vertices, and then draw the asymptotes and sketch the graph.

11. $25x^2 - 9y^2 = 225$

12. $y^2 - 4x^2 - 16 = 0$

13. $9y^2 - 4x^2 - 36 = 0$

14. $4x^2 - 9y^2 - 144 = 0$

15. $16x^2 - 9y^2 - 1 = 0$

Find the standard equation of each hyperbola whose graph is shown below.

16.

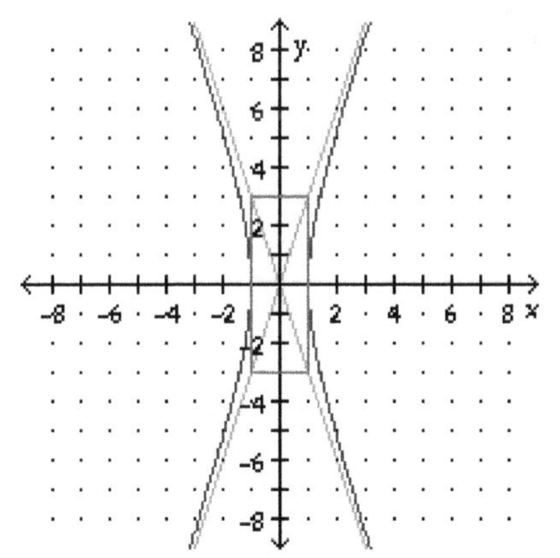

____

17.

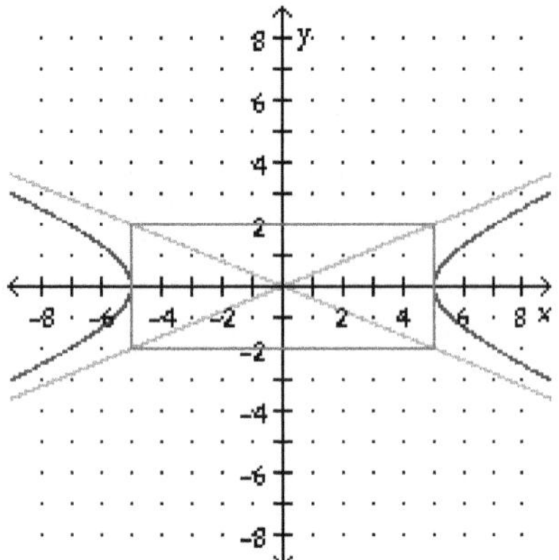
y
x
8
6
4
2
-2
-4
-6
-8
-8 -6 -4 -2 2 4 6 8

____

18.

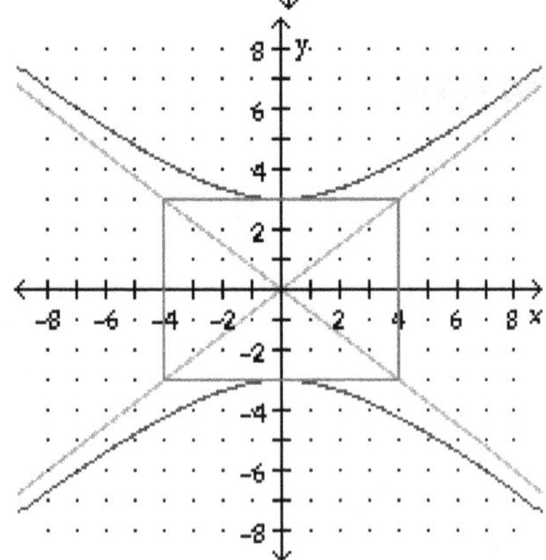
y
x
8
6
4
2
-2
-4
-6
-8
-8 -6 -4 -2 2 4 6 8

____

19.

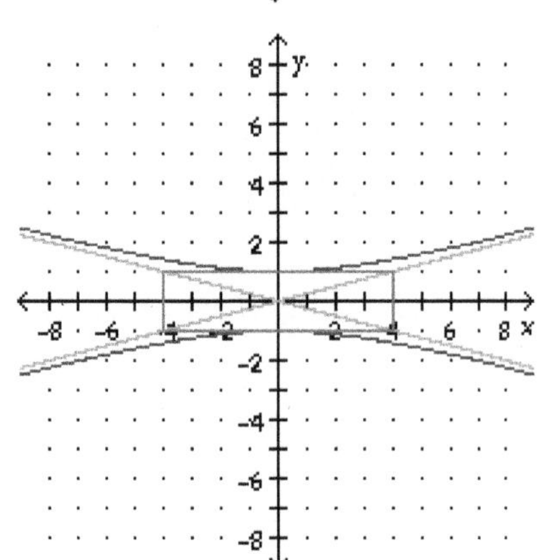
y
x
8
6
4
2
-2
-4
-6
-8
-8 -6 -4 -2 2 4 6 8

____

20.

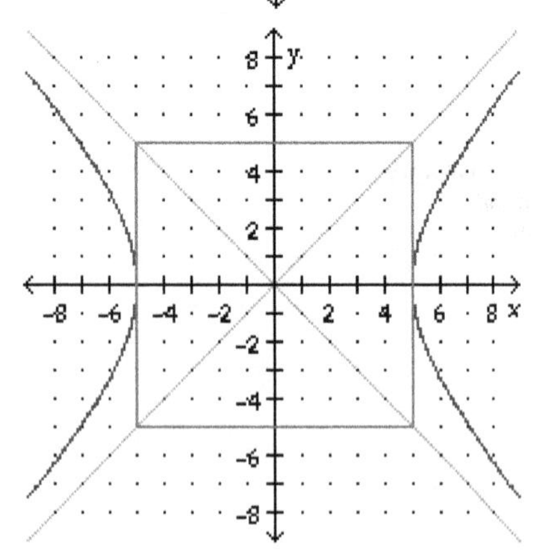
y
x
8
6
4
2
-2
-4
-6
-8
-8 -6 -4 -2 2 4 6 8

____

21.

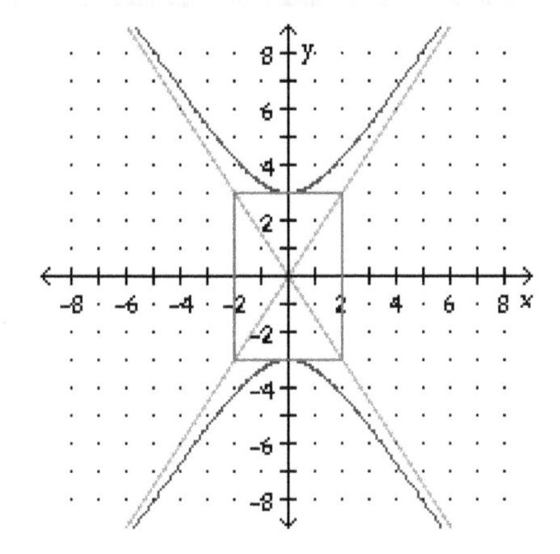

22.

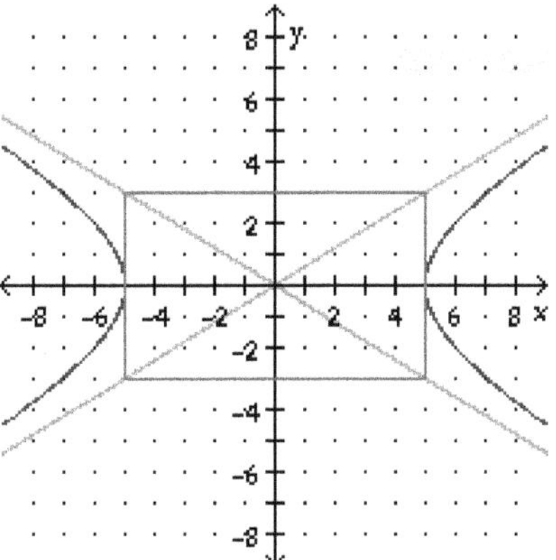

23.

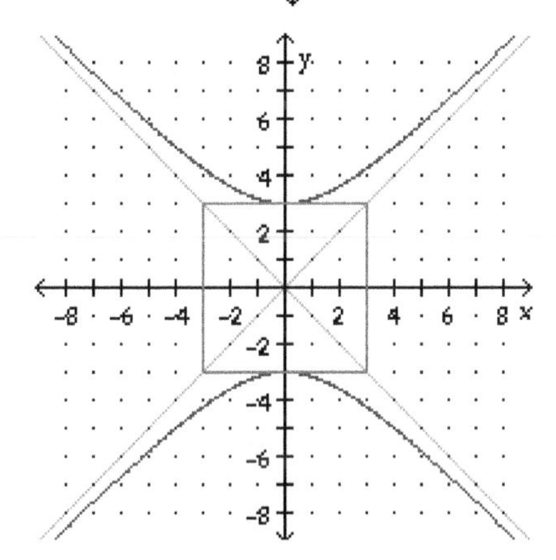

Find the standard equation of each hyperbola whose graph satisfies the conditions given below.

24. Vertices at $(1,0)$ and $(-1,0)$; foci at $\left(-\sqrt{5},0\right)$ and $\left(\sqrt{5},0\right)$

25. Vertices at $(0,3)$ and $(0,-3)$; foci at $\left(0,-\sqrt{13}\right)$ and $\left(0,\sqrt{13}\right)$

26. Vertices at $(2,0)$ and $(-2,0)$; foci at $\left(-2\sqrt{2},0\right)$ and $\left(2\sqrt{2},0\right)$

____

27. Vertices at $(0,-2)$ and $(0,2)$; foci at $\left(0,-\sqrt{11}\right)$ and $\left(0,\sqrt{11}\right)$

____

28. Vertices at $(6,0)$ and $(-6,0)$; asymptotes: $y=\pm\frac{3}{2}x$

____

29. Vertices at $\left(0,\sqrt{5}\right)$ and $\left(0,-\sqrt{5}\right)$; asymptotes: $y=\pm\frac{1}{5}x$

____

30. Vertices at $\left(\sqrt{3},0\right)$ and $\left(-\sqrt{3},0\right)$; asymptotes: $y=\pm 4x$

____

31. Vertices at $(0,4)$ and $(0,-4)$; asymptotes: $y=\pm\frac{2}{3}x$

____

32. Foci at $(5,0)$ and $(-5,0)$; asymptotes: $y=\pm\frac{3}{4}x$

____

33. Foci at $\left(0,-\sqrt{9}\right)$ and $\left(0,\sqrt{9}\right)$; asymptotes: $y=\pm\frac{\sqrt{5}}{2}x$

____

34. Foci at $(0,-4)$ and $(0,4)$; asymptotes: $y=\pm\frac{1}{\sqrt{3}}x$

____

35. Foci at $\left(5\sqrt{6},0\right)$ and $\left(-5\sqrt{6},0\right)$; asymptotes: $y=\pm\sqrt{5}x$

____

Chapter 7 Section 5: Translation of Conics

Practice Problems

Find the following for each parabola defined below: (a) the equation of the axis of symmetry; (b) the location of the vertex; (c) the equation of the directrix; and (d) the location of the focus. Use the information obtained about each parabola to sketch its graph.

1. $(x-1)^2 = 12(y+2)$

2. $(x+2)^2 = -(y+3)$

3. $(y-5)^2 = 6(x+6)$

4. $(y+3)^2 = -2(x-3)$

5. $y^2 = -8(x+4)$

Find the following for each ellipse defined below: (a) the direction of the major axis; (b) the location of the vertices; and (c) the location of the foci. Use the information obtained to sketch the graph.

6. $\dfrac{(x-3)^2}{4} + \dfrac{(y+2)^2}{25} = 1$

7. $\dfrac{(x+1)^2}{9} + \dfrac{(y+3)^2}{4} = 1$

8. $\dfrac{(x+4)^2}{25}+\dfrac{y^2}{16}=1$

9. $(x+6)^2+\dfrac{(y-5)^2}{4}=1$

10. $\dfrac{x^2}{49}+\dfrac{(y-3)^2}{36}=1$

Find the following for each hyperbola defined below: (a) the location of the vertices; (b) the location of the foci; (c) the direction of the transverse axis; and (d) the asymptotes' equations. Use the information obtained to sketch the graph.

11. $\dfrac{(x-3)^2}{16}-\dfrac{(y+2)^2}{16}=1$

12. $\dfrac{(y-4)^2}{25}-\dfrac{(x+5)^2}{9}=1$

13. $\dfrac{(x+6)^2}{4}-\dfrac{(y-1)^2}{49}=1$

14. $\dfrac{(y+3)^2}{9}-\dfrac{(x-6)^2}{25}=1$

15. $(x+2)^2-\dfrac{(y+1)^2}{36}=1$

Find the standard equation of each parabola whose graph is shown below.

16.

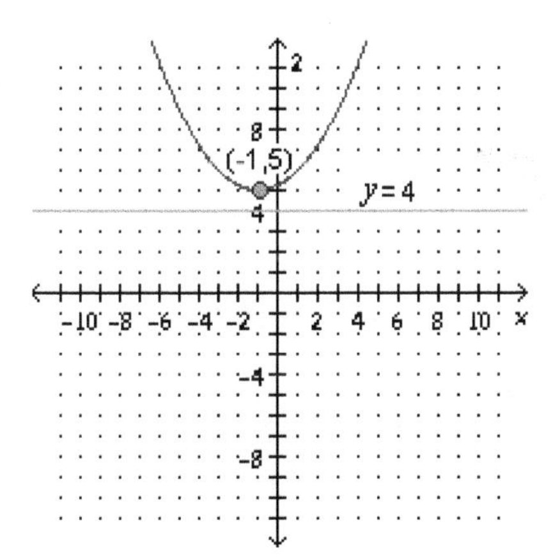

____

17.

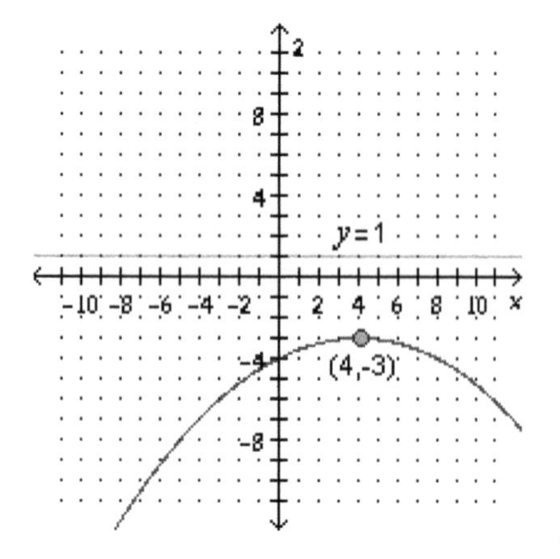

____

18.

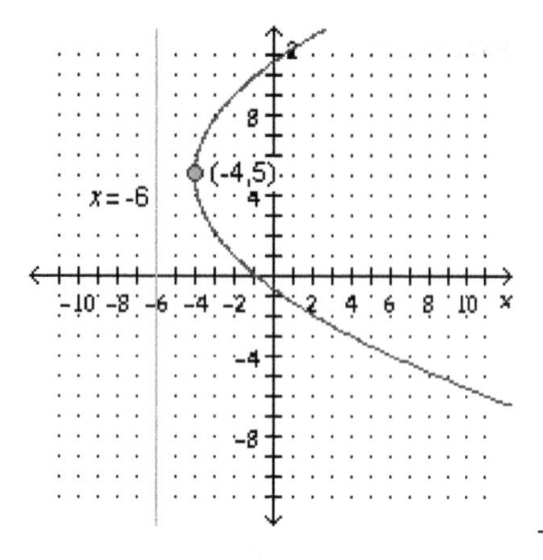

____

19.

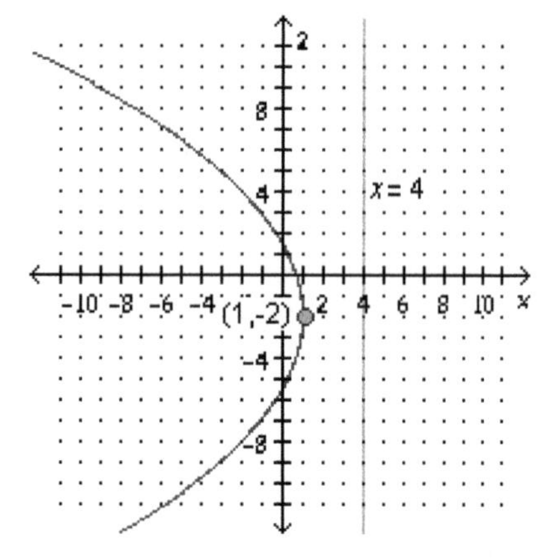

____

Find the standard equation of each ellipse whose graph is shown below.

20.

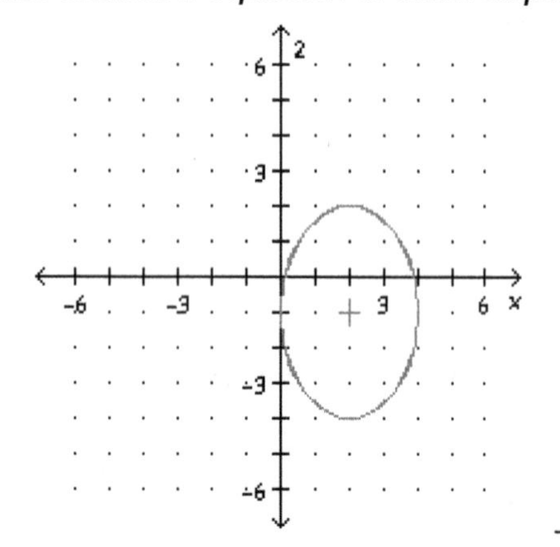

____

21.

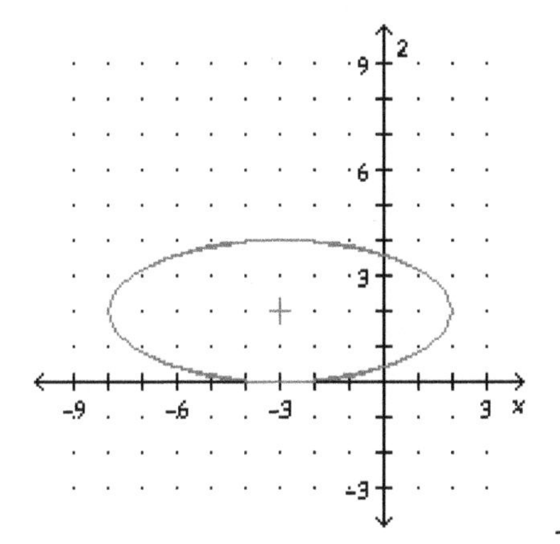

22.

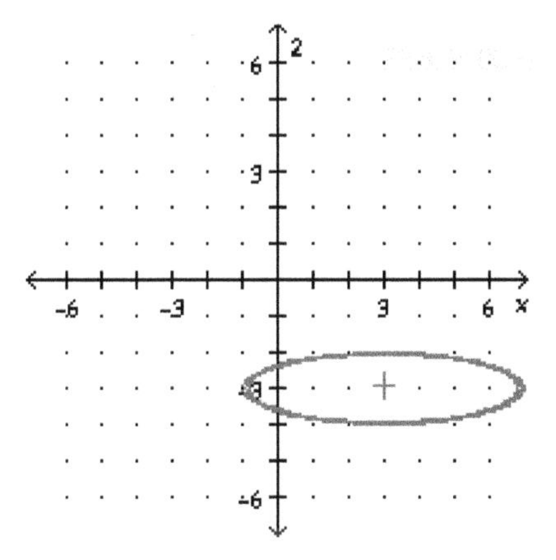

23.

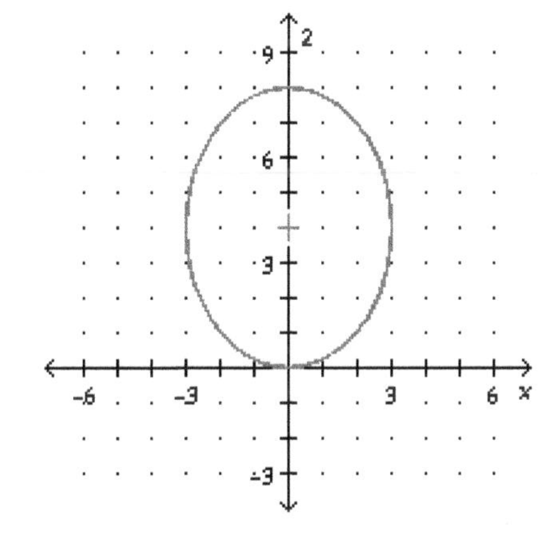

Find the standard equation of each hyperbola whose graph is shown below.

24.

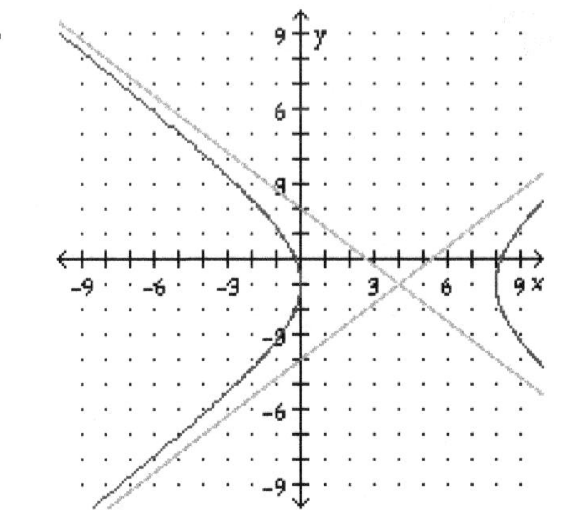

25.

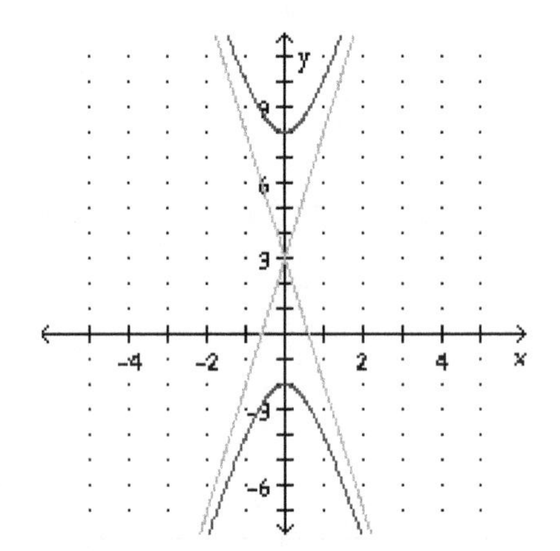

26.

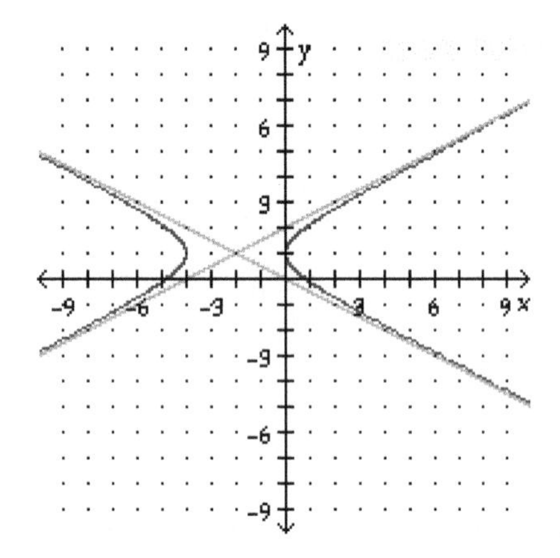

27.

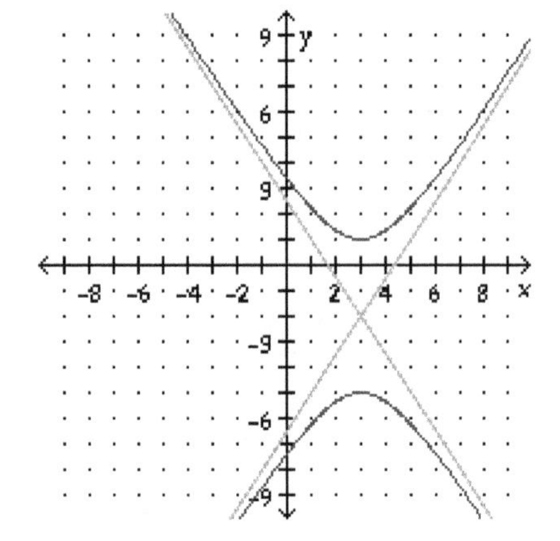

Transform the standard equation of each of the following conics by expanding its terms. Give an expression with no denominators and with all significant terms on the left of the equation. Decide whether the conic is a parabola, an ellipse, or a hyperbola. Also, state the direction and type of symmetry axis (major or transverse). (Answers may vary.)

28. $\frac{(x-1)^2}{4}+\frac{(y+3)^2}{9}=1$

29. $(x+5)^2=-8(y-3)$

30. 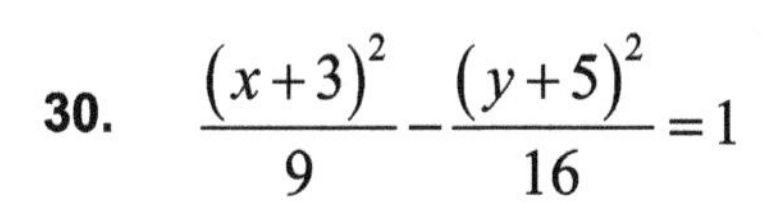

$$\frac{(x+3)^2}{9}-\frac{(y+5)^2}{16}=1$$

____

31. $(y+2)^2=\frac{4}{3}(x+4)$

____

32. $\frac{(x+3)^2}{16}+\frac{(y-3)^2}{25}=1$

____

33. $(y+4)^2-\frac{(x-3)^2}{36}=1$

____

Transform the general equation of each of the following conics into standard form. State which type of conic it represents.

34. $y^2-8x-2y+17=0$

____

35. $x^2+25y^2-14x+200y+424=0$

____

36. $4x^2-9y^2-32x+36y-8=0$

____

37. $-x^2+9y^2-10x-54y+47=0$

____

38. $x^2-16x+16y-48=0$

____

Find a polar equation for each indicated conic with the given characteristics.

39. Ellipse of eccentricity $\frac{1}{4}$, with focus at the origin and directrix on the line $x = 5$.

____

40. Parabola with focus at the origin and directrix on the line $y = -4$.

____

41. Hyperbola of eccentricity 4, with focus at the origin and directrix on the line $x = -7$.

____

42. Ellipse of eccentricity $\frac{2}{5}$, with focus at the origin and directrix on the line $y = \frac{1}{4}$.

____

43. Hyperbola of eccentricity 1.5, with focus at the origin and directrix on the line $y = -10$.

____

Identify the conic described by each of the following equations.

44. $r = \dfrac{5}{1 - 1.2\cos(\theta)}$ ____________________

45. $r = \dfrac{3}{2 + 2 \cdot \sin(\theta)}$ ____________________

46. $r = \dfrac{4}{5 - 3 \cdot \sin(\theta)}$ ____________________

47. $r = \dfrac{7}{2 + 6 \cdot \cos(\theta)}$ ____________________

Plot each of the following rotated conics.

48.

$$r = \frac{3}{5 + 4 \cdot \cos\left(\theta - \frac{\pi}{3}\right)}$$

49. $$r = \frac{4}{6 - 5 \cdot \cos\left(\theta - \frac{3\pi}{4}\right)}$$

50. $$r = \frac{2}{5 + 3 \cdot \sin\left(\theta - \frac{5\pi}{6}\right)}$$

51. $$r = \frac{5}{6 - 4 \cdot \sin\left(\theta - \frac{\pi}{6}\right)}$$

52. $$r = \frac{8}{6 + 5 \cdot \cos\left(\theta + \frac{5\pi}{6}\right)}$$

CONTENT on Demand

PRECALCULUS 2ND EDITION

Practice Problem Worksheets

CHAPTER 8: SYSTEMS OF EQUATIONS

Practice Problems

Solve each of the following systems of equations by graphing.

1. $\begin{Bmatrix} 2x-2y=16 \\ 3x-y=14 \end{Bmatrix}$

2. $\begin{Bmatrix} 5x+2y=-22 \\ 3x-4y=18 \end{Bmatrix}$

3. $\begin{Bmatrix} 2x-4y=-18 \\ x+3y=26 \end{Bmatrix}$

4. $\begin{cases} 5x+3y=4 \\ -2x-3y=-16 \end{cases}$

5. $\begin{cases} -2x+3y=26 \\ 4x+3y=-16 \end{cases}$

Use substitution to solve each system of equations below.

6. $\begin{cases} 2x-3y=9 \\ -4x-y=-4 \end{cases}$ ____________

7. $\begin{cases} -2x-5y=9 \\ 5x+4y=3 \end{cases}$ ____________

8. $\begin{cases} 3x-4y=-13 \\ 5x+2y=-26 \end{cases}$ ____________

9. $\begin{cases} 4x-2y=-8 \\ 2x+3y=0 \end{cases}$ ____________

10. $\begin{cases} 5x+3y=25 \\ 3x-7y=-7 \end{cases}$ ____________

Solve each system of equations below by using the elimination method.

11. $\begin{cases} 3x-2y=-3 \\ 6x+y=4 \end{cases}$ ____________

12. $\begin{cases} 4x-y=-15 \\ 3x-2y=-10 \end{cases}$ ____________

13. $\begin{cases} 2x+3y=0 \\ 6x-5y=-28 \end{cases}$ ____________

14. $\begin{cases} 8x+5y=29 \\ 2x-4y=-19 \end{cases}$ ____________

15. $\begin{cases} 3x-2y=20 \\ 2x+3y=-17 \end{cases}$ ____________

16. $\begin{cases} 5x-4y=-52 \\ 3x+3y=12 \end{cases}$ ____________

*Decide whether each system of equations will render **one solution** (lines that intersect), **no solution** (parallel lines), or **infinite number** of solutions (overlapping lines).*

17. $\begin{cases} 4x+5y=37 \\ -2x+3y=9 \end{cases}$ ____________

18. $\begin{cases} 2x-7y=2 \\ -6x+21y=15 \end{cases}$ ____________

19. $\begin{cases} 15x-20y=25 \\ -18x+24y=-30 \end{cases}$ ____________

20. $\begin{cases} -12x+4y=-32 \\ 9x-3y=26 \end{cases}$ ____________

21. $\begin{cases} -3x+2y=-21 \\ 5x-4y=26 \end{cases}$ ____________

22. The difference of two numbers is 17. Find the numbers if their sum is 87.

____

23. One number is 9 times a second number and their sum is 120. Find the numbers.

____

24. Sally is three times as old as Nick. The sum of their ages is 24 years. How old are they?

____

25. A box contains dimes and nickels totaling $7.35. If there are 12 more dimes than nickels, how many of each type are in the box?

____

26. A coin collector has a certain number of coins worth $12.50 each, and a second type of coin worth $2.75 each. He has ten fewer coins of the first kind than of the second kind. The total value of the coins is $119. How many coins of each type does he have?

____

27. A person sells two types of jerky. He sells twice as many beef jerky bas as turkey jerky bags. His profit from the beef jerky is $1.50 per bag, while for the turkey jerky it is $2.00 per bag. How many bags of each type must he sell to make a $600 profit?

____

28. A chemist needs to mix a 10% acid solution with a 25% acid solution to produce 12 liters of a 15% acid mixture. How much of each type solution should he mix?

____

29. How many gallons of 20% weed killer should be added to 30 gallons of a 50% weed killer solution to obtain a 40% weed killer solution?

____

30. How many grams of an alloy that is 70% gold should be melted with 60 grams of an alloy that is 20% gold to produce an alloy that is 30% gold?

____

31. A person invested part of $20,000 at 8% interest and the rest at 7%. If her income from the 8% investment is $250 more than the income from the 7% investment, how much was invested at each rate?

____

32. A person invested $25,000 in the stock market. Investment A gave him 18% profit while in investment B he lost 11%. Knowing that his net gain was $2180, how much was placed in A and how much in B?

____

Chapter 8 Section 2: Larger Systems of Linear Equations

Practice Problems

Solve each system of equations below by using the elimination method.

1. $$\begin{cases} 2x+3y-2z=-3 \\ 3x-6z=-33 \\ x-2y+4z=13 \end{cases}$$ ____________________

2. $$\begin{cases} 2x-2y-3z=8 \\ -5x+3y-7z=-16 \\ -3x+4y+z=-14 \end{cases}$$ ____________________

3. $$\begin{cases} 4x+3y+2z=-9 \\ -x+2y+3z=17 \\ -3x-4y-2z=6 \end{cases}$$ ____________________

4. $$\begin{cases} 3x+4y-2z=-15 \\ -2x+3y+z=-29 \\ 2x-y+8z=-3 \end{cases}$$ ____________________

5. $$\begin{cases} -2x+3y-z=-16 \\ -x-2y+4z=15 \\ 6x-9y+3z=4 \end{cases}$$ ____________________

Write each system of equations below in augmented matrix form. Do not solve the system.

6. $$\begin{cases} -2x+3y-z=-1 \\ -x-2z+3y=8 \\ 4y-3z+3x=-10 \end{cases}$$ ____________________

7. $$\begin{cases} -4x+2y=9 \\ x+5z=10 \\ 2x-y+4z=0 \end{cases}$$ ____________________

8. $$\begin{cases} -2x=+3y-z \\ -x-4y=7z-2 \\ 3y=2-3x \end{cases}$$ ____________________

9. $\begin{Bmatrix} -3+4x=6y-z \\ -1=-2z \\ 4x-3z+4y-1=0 \end{Bmatrix}$

____

10. $\begin{Bmatrix} -3y+3=6z \\ 4-x+3z=-2y \\ 6-5z+x=-3y \end{Bmatrix}$

____

Solve each system of equations by converting it into an augmented matrix. Perform the appropriate row operations to reduce it to echelon form.

11. $\begin{Bmatrix} 4x-y+3z=-34 \\ -2x+2y+z=8 \\ x+3y+2z=-3 \end{Bmatrix}$

____

12. $\begin{Bmatrix} 2x+5y-2z=1 \\ -x-3y+4z=12 \\ 4y+3z=8 \end{Bmatrix}$

____

13. $\begin{Bmatrix} -6x+5y-3z=-2 \\ 2x+4y-z=11 \\ 5x-3y+2z=3 \end{Bmatrix}$

____

14. $\begin{Bmatrix} 2x-3y+4z=18 \\ -3x-2y=8 \\ 5x+3z=4 \end{Bmatrix}$

____

15. $\begin{Bmatrix} 2x - y + 3z = 21 \\ -x + 3y + 2z = -27 \\ 3x + 4y + 6z = -11 \end{Bmatrix}$

____

Decide whether each matrix below is in echelon form (EF) or reduced row-echelon form (RREF) or neither. Justify your answer.

16. $\begin{bmatrix} 1 & 0 & 2 & 4 \\ 0 & 0 & 1 & -3 \\ 0 & 1 & 0 & 5 \end{bmatrix}$ ______________________________

17. $\begin{bmatrix} 1 & 0 & 0 & 8 \\ 0 & 1 & 0 & 4 \\ 0 & 0 & 1 & -6 \end{bmatrix}$ ______________________________

18. $\begin{bmatrix} 1 & -1 & 0 & 7 \\ 0 & 1 & 0 & 6 \\ 0 & 0 & 1 & -2 \end{bmatrix}$ ______________________________

19. $\begin{bmatrix} 1 & 0 & 0 & 3 \\ 0 & 0 & 0 & 0 \\ 0 & 0 & 1 & 1 \end{bmatrix}$ ______________________________

20. $\begin{bmatrix} 1 & 0 & 0 & -3 \\ 0 & 1 & 5 & -2 \\ 0 & 0 & 0 & 0 \end{bmatrix}$ ______________________________

21. $\begin{bmatrix} 1 & 3 & 0 & 0 \\ 0 & 0 & 1 & 0 \\ 0 & 0 & 0 & 1 \end{bmatrix}$ ______________________________

22. $\begin{bmatrix} 1 & -6 & 2 & 1 \\ 0 & 1 & 1 & 0 \\ 0 & 0 & 1 & 6 \end{bmatrix}$ ______________________________

23. 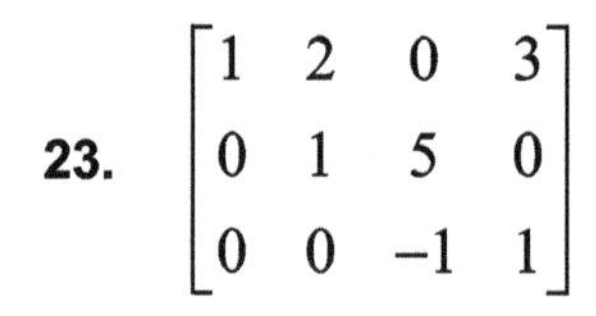

$$\begin{bmatrix} 1 & 2 & 0 & 3 \\ 0 & 1 & 5 & 0 \\ 0 & 0 & -1 & 1 \end{bmatrix}$$

____

24. $$\begin{bmatrix} 1 & 0 & 0 & -7 \\ 0 & 1 & -2 & 4 \\ 0 & 1 & 0 & 0 \end{bmatrix}$$

____

25. $$\begin{bmatrix} 1 & 4 & 8 & 9 \\ 0 & 1 & -6 & 3 \\ 0 & 0 & 0 & 0 \end{bmatrix}$$

____

Write the linear equation associated with each matrix given below.

26. $$\begin{bmatrix} 1 & -3 & 0 & 8 \\ 0 & 1 & -2 & 7 \\ 0 & 0 & 0 & 0 \end{bmatrix}$$

____

27. $$\begin{bmatrix} 1 & -1 & 4 & -3 \\ 0 & 1 & -5 & 8 \\ 0 & 0 & 1 & -6 \end{bmatrix}$$

____

28. $$\begin{bmatrix} 1 & 0 & 4 & -5 \\ 0 & 1 & -3 & 10 \\ 0 & 0 & 0 & 0 \end{bmatrix}$$

____

29. $$\begin{bmatrix} 1 & 5 & -8 & 11 \\ 0 & 1 & 9 & -3 \\ 0 & 0 & 1 & 4 \end{bmatrix}$$

____

30. $$\begin{bmatrix} 1 & 0 & -4 & -5 \\ 0 & 1 & 8 & -9 \\ 0 & 0 & 0 & 0 \end{bmatrix}$$

____

Decide on the number and type of solutions to each system of equations whose matrices are given below in reduced row-echelon form. Justify your answer.

31. $$\begin{bmatrix} 1 & 0 & 0 & -7 \\ 0 & 1 & 0 & 4 \\ 0 & 0 & 1 & -2 \end{bmatrix}$$

____

32. $\begin{bmatrix} 1 & 0 & 0 & 5 \\ 0 & 1 & 0 & 0 \\ 0 & 0 & 1 & 0 \end{bmatrix}$

____

33. $\begin{bmatrix} 1 & 0 & 0 & 4 \\ 0 & 1 & 0 & -10 \\ 0 & 0 & 0 & 0 \end{bmatrix}$

____

34. $\begin{bmatrix} 1 & 2 & 0 & -3 \\ 0 & 0 & 1 & 4 \\ 0 & 0 & 0 & 0 \end{bmatrix}$

____

35. $\begin{bmatrix} 1 & 0 & 0 & -6 \\ 0 & 0 & 1 & 5 \\ 0 & 0 & 0 & 1 \end{bmatrix}$

____

Practice Problems

Use the "Rref" option of the Matrix tool in Calculator on Demand to solve each system of equations and answer the questions below.

1. $\begin{cases} 3x-2y+3z=17 \\ -2x-2y+4z=26 \\ 7x+3y-2z=-3 \end{cases}$ ____________________________

2. $\begin{cases} 4x-y+3z=-34 \\ -2x+2y+z=8 \\ x+3y+2z=-3 \end{cases}$ ____________________________

3. $\begin{cases} 2x-y+3z=21 \\ -x+3y+2z=-27 \\ 3x+4y+6z=-11 \end{cases}$ ____________________________

4. $\begin{cases} -3x-2y=8 \\ 2x-3y+4z=18 \\ 5x+3z=4 \end{cases}$ ____________________________

5. $\begin{cases} 5x-4y-2z=31 \\ 4x-5y=33 \\ -2x-3y+4z=5 \end{cases}$ ____________________________

6. A person has in his pocket nickels, dimes and quarters worth $7.50. He has 3 fewer dimes than nickels and twice as many quarters as nickels. How many nickels does he have?

____

7. The following table shows the number of calories, milligrams of sodium and grams of fat contained in one ounce of the following nuts (as stated by the producer).

	almonds	hazelnuts	peanuts
Calories	165	150	120
Sodium	110	90	80
Fat	5	6	8

The three nuts are to be combined in a trail mix that provides climbers with a serving of 1095 calories, 700 milligrams of sodium and 55 grams of fat. How many ounces of hazelnut should be used in the mix?

____

8. A cookie factory sells special cookie packages for Christmas. Package type A contains 0.6 of a pound of ginger cookies and 0.4 of a pound of coconut cookies. Package type B contains 0.5 of a pound of ginger cookies, 0.3 of a pound of coconut cookies and 0.2 of a pound of chocolate cookies. Package type C contains 0.3 of a pound of ginger cookies, 0.4 of a pound of coconut cookies and 0.3 of a pound of chocolate cookies. The company has produced 207 pounds of ginger cookies, 147 pounds of coconut cookies and 81 pounds of chocolate cookies. How many packages of type "C" should they make to use up all the cookies they produced?

____

Practice Problems

Four matrices of dimension 2x3 are given below. Perform each indicated matrix operation.

$$A=\begin{bmatrix}3 & -2 & 6\\ 5 & 0 & -1\end{bmatrix} \quad B=\begin{bmatrix}4 & -7 & 2\\ 2 & 6 & -3\end{bmatrix} \quad C=\begin{bmatrix}-5 & 8 & 0\\ 3 & -4 & 2\end{bmatrix} \quad D=\begin{bmatrix}3 & 7 & 2\\ -4 & 8 & 9\end{bmatrix}$$

1. A + B ____________________

2. A - C ____________________

3. C + D + A ____________________

4. B + C - D ____________________

5. 2C - A ____________________

6. 4B + 3C ____________________

7. 2B - 5A ____________________

8. 3C - 2D ____________________

Five matrices of various dimensions are given below. Decide if each indicated matrix multiplication is possible and, if so, give the resulting product matrix.

$$A=\begin{bmatrix}1 & 3 & 0 & -2\\ 5 & 2 & -1 & 1\\ 0 & 3 & 2 & 6\end{bmatrix} \quad B=\begin{bmatrix}2 & 4 & -1\\ 0 & 3 & 5\\ -2 & 1 & -3\\ 4 & -2 & 0\end{bmatrix} \quad C=\begin{bmatrix}6 & -1 & 0\\ 2 & 3 & 5\\ -1 & 4 & 2\end{bmatrix}$$

$$D=\begin{bmatrix}3 & 5 & -1\\ 2 & 0 & -4\\ 1 & 1 & 0\end{bmatrix} \quad E=\begin{bmatrix}3 & -2 & 0 & 1\\ 4 & 2 & 1 & 5\\ 0 & -3 & -1 & 2\\ 1 & 0 & -2 & 4\end{bmatrix}$$

9. $A \cdot B$

____

10. $B \cdot A$

____

11. $A \cdot C$

____

12. $C \cdot B$

____

13. $D \cdot A$

____

14. $E \cdot A$

____

$$A=\begin{bmatrix}1 & 3 & 0 & -2\\5 & 2 & -1 & 1\\0 & 3 & 2 & 6\end{bmatrix}\qquad B=\begin{bmatrix}2 & 4 & -1\\0 & 3 & 5\\-2 & 1 & -3\\4 & -2 & 0\end{bmatrix}\qquad C=\begin{bmatrix}6 & -1 & 0\\2 & 3 & 5\\-1 & 4 & 2\end{bmatrix}$$

$$D=\begin{bmatrix}3 & 5 & -1\\2 & 0 & -4\\1 & 1 & 0\end{bmatrix}\qquad E=\begin{bmatrix}3 & -2 & 0 & 1\\4 & 2 & 1 & 5\\0 & -3 & -1 & 2\\1 & 0 & -2 & 4\end{bmatrix}$$

15. $E \cdot B$

____

16. $A \cdot E$

____

17. $D \cdot E$

____

18. $C \cdot D$

____

Find the inverse (if it exists) of each matrix given below.

19. $\begin{bmatrix}3 & -1 & 0\\0 & 2 & 4\\1 & 3 & -1\end{bmatrix}$

____

20. $\begin{bmatrix}2 & 4 & -2\\0 & 3 & -1\\1 & 0 & 2\end{bmatrix}$

____

21. 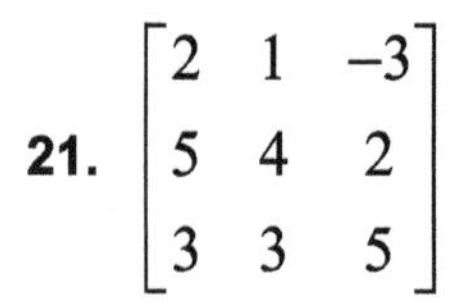

$$\begin{bmatrix} 2 & 1 & -3 \\ 5 & 4 & 2 \\ 3 & 3 & 5 \end{bmatrix}$$

____

22. $\begin{bmatrix} 1 & 0 & -2 \\ 3 & -1 & 4 \\ 2 & -1 & 6 \end{bmatrix}$

____

23. $\begin{bmatrix} 3 & 4 & -2 \\ 1 & 0 & 7 \\ -2 & 5 & 0 \end{bmatrix}$

____

Write each system of equations below as a matrix equation. Do not solve the system.

24. $\begin{Bmatrix} 3x-2y+z=-3 \\ -x+y-2z=0 \\ 2x-2y+4z=6 \end{Bmatrix}$ ______________________________

25. $\begin{Bmatrix} 3y-7z=10 \\ 5x-2y+z=-8 \\ -2x+y-3z=-15 \end{Bmatrix}$ ______________________________

26. $\begin{Bmatrix} x+y-z=5 \\ -2x-y-3z=-4 \\ 4x-2z=13 \end{Bmatrix}$ ______________________________

27. $\begin{Bmatrix} -3x+5y-3z=2 \\ -x+y+9z=3 \\ 4x+3y=12 \end{Bmatrix}$ ______________________________

28. $\begin{Bmatrix} 2y-5z=-18 \\ -3x+5y+z=7 \\ 3x-4y+2z=16 \end{Bmatrix}$ ______________________________

Solve each system of equations by writing it as a matrix equation and finding the inverse.

29. $\begin{cases} -x+3y+2z=7 \\ 2x+y-2z=-10 \\ -4x-5y=12 \end{cases}$ ____________________

30. $\begin{cases} 4x-2y-z=3 \\ -2x+3y+2z=-5 \\ 2y+z=-1 \end{cases}$ ____________________

31. $\begin{cases} 2x-3y-2z=11 \\ x+2y+3z=18 \\ -x-2y+5z=14 \end{cases}$ ____________________

32. $\begin{cases} 2x-y+z=11 \\ 4x+2y+2z=30 \\ x-5y+z=-7 \end{cases}$ ____________________

33. $\begin{cases} -5x-y+2z=6 \\ x-4y-2z=-20 \\ 2x+y+4z=-18 \end{cases}$ ____________________

Practice Problems

Solve each system of linear equations algebraically. Try to visualize the solutions graphically by analyzing the type of equations involved in the system.

1. $\begin{cases} x^2 + y^2 = 8 \\ x \cdot y = 4 \end{cases}$ ________________________

2. $\begin{cases} x^2 + y^2 = 4 \\ x^2 - 6x + y^2 = 0 \end{cases}$ ________________________

3. $\begin{cases} x^2 + y^2 = 6 \\ y^2 = x \end{cases}$ ________________________

4. $\begin{cases} x^2 + y^2 = 16 \\ 9x^2 + 25y^2 = 225 \end{cases}$ ________________________

5. $\begin{cases} 9x^2 + 4y^2 = 36 \\ y^2 - 4x = 8 \end{cases}$ ________________________

6. $\begin{cases} x^2 + y = 4 \\ 6x^2 - 3y = -12 \end{cases}$ ________________________

7. $\begin{cases} -x^2 + y = 6 \\ 9x^2 + y^2 = 9 \end{cases}$ ________________________

8. $\begin{cases} 4x^2 + 25y^2 = 100 \\ x + y = 2 \end{cases}$ ________________________

9. $\begin{cases} xy - 4(x + y) = -14 \\ x \cdot y = 2 \end{cases}$ ________________________

10. $\begin{cases} 4y^2 - x^2 = 4 \\ y - 2x = 0 \end{cases}$ ________________________

Solve each system of nonlinear equations given in function form. In some cases you may need to use a graphing calculator since there is no simple algebraic method to solve the resulting equation.

11. $\begin{cases} y = x^3 + 2x^2 - 3x + 2 \\ y = x^3 + x^2 + 2x + 16 \end{cases}$ ________________

12. $\begin{cases} y = 2x^4 - x^3 + 2x^2 - x - 12 \\ y = x^4 - x^3 - x + 12 \end{cases}$ ________________

13. $\begin{cases} y = \sqrt{-x^2 + 6x + 1} \\ y = \sqrt{x^2 - 4x + 13} \end{cases}$ ________________

14. $\begin{cases} x \cdot y = -12 \\ y = x^2 - 3x - 4 \end{cases}$ ________________

15. $\begin{cases} y = 2x^3 - 7x^2 - 5x - 11 \\ y = -4x^3 + 18x^2 - 2x + 16 \end{cases}$ ________________

16. $\begin{cases} y = \sqrt{21 - x^2 - 4x} \\ y = -x^3 - 15x^2 - 75x - 121 \end{cases}$ ________________

CONTENT on Demand

Precalculus 2nd Edition

Practice Problem Worksheets

Chapter 9: Sequences & Series

Chapter 9 Section 1: Sequences and Series–Generalities and Notation

Practice Problems

Find the first five terms of each sequence defined by the given n^{th} term.

1. $a_n = 3n - 1$ ____________________

2. $a_n = -2$ ____________________

3. $a_n = \dfrac{2}{n+2}$ ____________________

4. $a_n = n^2 + 2$ ____________________

5. $a_n = n^2 - n$ ____________________

6. $a_n = \dfrac{(-1)^n}{n+1}$ ____________________

7. $a_n = \dfrac{(-1)^{n+1}}{n^2+2}$ ____________________

8. $a_n = \dfrac{(-1)^n}{3n}$ ____________________

9. $a_n = \dfrac{(-2)^n}{n}$ ____________________

10. $a_n = \dfrac{2^n}{n^2+n}$ ____________________

Find the nth term of each sequence whose first few terms are given.

11. $\dfrac{2}{1}, \dfrac{3}{2}, \dfrac{4}{3}, \dfrac{5}{4}, \ldots$ ____________

12. $-3, 3, -3, 3, \ldots$ ____________

13. $0, \dfrac{1}{4}, \dfrac{2}{5}, \dfrac{1}{2}, \ldots$ ____________

14. $1, \dfrac{1}{4}, \dfrac{1}{16}, \dfrac{1}{64}, \ldots$ ____________

15. $4, -8, 16, -32, \ldots$ ____________

16. $1, 5, 9, 13, \ldots$ ________________

17. $\frac{1}{4}, -\frac{1}{6}, \frac{1}{8}, -\frac{1}{10}, \ldots$ ________________

18. $2, -4, 8, -16, \ldots$ ________________

19. $\frac{1}{2}, \frac{2}{5}, \frac{3}{10}, \frac{4}{17}, \ldots$ ________________

20. $\frac{1}{4}, 0, \frac{1}{4}, 0, \ldots$ ________________

Find the first four terms of each sequence defined below.

21. $a_n = 2a_{n-1} + 3$; with $a_1 = 5$

____

22. $a_n = \frac{a_{n-1} + 1}{2}$; with $a_1 = -3$

____

23. $a_n = (a_{n-1})^2 + 2$; with $a_1 = 2$

____

24. $a_n = \frac{3 - a_{n-1}}{2}$; with $a_1 = 5$

____

25. $a_n = \dfrac{a_{n-1} + 3a_{n-2}}{2}$; with $a_1 = -2$ and $a_2 = 1$

____

Find the partial sums S_1, S_2, S_3, S_4 and S_5 of each given sequence.

26. $2, -1, 2, -1, \ldots$

____

27. $1, 2, 4, 8, \ldots$

____

28. $3, 7, 11, 15, \ldots$

____

29. $\frac{1}{2}, \frac{1}{4}, \frac{1}{6}, \frac{1}{8}, \ldots$

____

30. $1, -\frac{1}{2}, \frac{1}{3}, -\frac{1}{4}, \ldots$

31. $a_n = \sqrt{2n} - \sqrt{2n+2}$

32. $a_n = \frac{1}{n} - \frac{1}{n+1}$

33. $a_n = \frac{2}{n+3} - \frac{2}{n+4}$

Find each indicated sum.

34. $\sum_{k=1}^{6} 5k$ ________________________________

35. $\sum_{k=3}^{10} 4$ ________________________________

36. $\sum_{k=2}^{4} \frac{2}{3^{k-1}}$ ________________________________

37. $\sum_{k=2}^{5} \frac{(-1)^k}{2k}$ ________________________________

38. $\sum_{k=2}^{6} 2k-3$ ________________________________

39. $\sum_{k=3}^{6} \frac{2k+1}{k-1}$ ________________________________

40. $\sum_{k=3}^{7} 2^{k-2}$ ________________________________

41. $\sum_{k=2}^{6} 3\left(\frac{1}{k-1}-\frac{1}{k}\right)$ ________________________________

Chapter 9 Section 2: Arithmetic Sequences and Series

Practice Problems

Decide if each given sequence is an arithmetic sequence and, if so, give the common difference (d).

1. $1, 4, 7, 10, \ldots$ ____________________

2. $2, 7, 9, 13, \ldots$ ____________________

3. $1, 3, 9, 11 \ldots$ ____________________

4. $3, -1, -5, -9, \ldots$ ____________________

5. $\frac{2}{5}, \frac{1}{5}, 0, -\frac{1}{5}, \ldots$ ____________________

6. $2, \frac{5}{2}, 3, \frac{7}{2}, \ldots$ ____________________

7. $-2, -4, -8, -16, \ldots$ ____________________

8. $\frac{1}{8}, 0, \frac{1}{8}, 0, \ldots$ ____________________

9. $4, \frac{15}{4}, \frac{7}{2}, \frac{13}{4}, \ldots$ ____________________

10. $\frac{1}{2}, \frac{1}{3}, \frac{1}{4}, \frac{1}{5}, \ldots$ ____________________

11. $\frac{1}{2}, \frac{1}{7}, \frac{1}{12}, \frac{1}{17}, \ldots$ ____________________

12. $a_n = 2n + 5$ ____________________

13. $a_n = -n + 4$ ____________________

14. $a_n = n^2 - 2$ ____________________

15. $a_n = 3n - 7$ ____________________

Find the indicated term for each given sequence.

16. 51st term in the sequence $-2, 6, 14, \ldots$ ________________

17. 28th term in the sequence $\frac{2}{3}, \frac{1}{3}, 0, \ldots$ ________________

18. 32nd term in the sequence $7, -2, -11, \ldots$ ________________

19. 64th term in the sequence $-32, -29, -26, \ldots$ ________________

20. 45th term in the sequence $\frac{3}{8}, \frac{5}{8}, \frac{7}{8}, \ldots$ ________________

21. 73rd term in the sequence $10, 10.2, 10.4, \ldots$ ________________

22. 83rd term in the sequence $-50, -43, -36, \ldots$ ________________

23. 91st term in the sequence $31, 28, 25, \ldots$ ________________

24. 37th term in the sequence defined by $a_n = 10 - 4n$ ________________

25. 42nd term in the sequence defined by $a_n = \frac{n}{2} + 5$ ________________

Find the indicated partial sum for each arithmetic sequence below.

26. S_{12} for the sequence with $a_1 = 3$ and $d = 4$. ________________

27. S_{15} for the sequence with $a_1 = 72$ and $d = -7$. ________________

28. S_{22} for the sequence with $a_1 = \frac{1}{2}$ and $d = \frac{3}{4}$. ________________

29. S_{18} for the sequence with $a_1 = -5$ and $d = \frac{1}{2}$. ________________

30. S_{25} for the sequence with $a_1 = \frac{3}{4}$ and $d = -\frac{1}{8}$. ________________

Evaluate the indicated partial sum of each arithmetic sequence.

31. $\sum_{k=1}^{20} 4k-3$ ________________

32. $\sum_{k=5}^{32} 2-3k$ ________________

33. $\sum_{k=7}^{21} 5k+2$ ________________

34. $\sum_{k=6}^{35} \frac{k}{4}+2$ ________________

35. $\sum_{k=4}^{43} 1-\frac{k}{10}$ ________________

Find the first term of each arithmetic sequence with the given characteristics.

36. Common difference $d=\frac{5}{2}$ and $a_{12}=\frac{35}{2}$ ________________

37. Common difference $d=-\frac{3}{2}$ and $a_{25}=-29$ ________________

38. Common difference $d=3$ and $a_{10}=\frac{45}{2}$ ________________

39. Common difference $d=0.1$ and $a_{18}=2.2$ ________________

40. Common difference $d=-\frac{7}{2}$ and $a_{20}=-\frac{91}{2}$ ________________

Find the indicated sum for each arithmetic sequence with the given characteristics.

41. S_{20} for the sequence with $a_1=3$ and $d=-1.2$ ________________

42. S_{18} for the sequence with $a_1 = -25$ and $d = 4$

____

43. S_{22} for the sequence with $a_1 = \frac{3}{2}$ and $d = \frac{7}{2}$

____

44. S_{25} for the sequence with $a_1 = 0.1$ and $d = -0.2$

____

45. S_{17} for the sequence with $a_1 = 42$ and $d = -5.5$

____

Find the indicated sum for each arithmetic sequence with the specified terms.

46. S_{15} for the sequence with $a_2 = 0.35$ and $a_{10} = 1.55$

____

47. S_{12} for the sequence with $a_3 = -17$ and $a_9 = 7$

____

48. S_{20} for the sequence with $a_4 = 12$ and $a_{15} = 50.5$

____

49. S_{14} for the sequence with $a_5 = -0.7$ and $a_{12} = -2.1$

____

50. S_{18} for the sequence with $a_6 = 62$ and $a_{16} = -8$

____

Chapter 9 Section 3: Geometric Sequences and Series

Practice Problems

Find the first four terms of each geometric sequence whose first term and common ratio are given.

1. $a_1 = 0.03$ and $r = 2$ ____________________

2. $a_1 = 2500$ and $r = \frac{1}{10}$ ____________________

3. $a_1 = 768$ and $r = \frac{1}{2}$ ____________________

4. $a_1 = 400$ and $r = \frac{1}{5}$ ____________________

5. $a_1 = \frac{4}{9}$ and $r = 3$ ____________________

Decide if each given sequence is a geometric sequence and, if so, give the common ratio (r).

6. $\frac{1}{2}, \frac{1}{6}, \frac{1}{18}, \ldots$ __________

7. $-3, 1, 3, \ldots$ __________

8. $-\frac{1}{8}, -\frac{1}{2}, -2, \ldots$ __________

9. $-\frac{2}{5}, \frac{1}{5}, -\frac{2}{5}, \ldots$ __________

10. $\frac{1}{4}, -\frac{1}{8}, \frac{1}{64}, \ldots$ __________

11. $2, \frac{10}{3}, \frac{50}{9}, \ldots$ __________

12. $-\frac{1}{5}, \frac{3}{2}, -\frac{1}{10}, \ldots$ __________

13. $\frac{2}{3}, \frac{3}{2}, \frac{2}{3}, \ldots$ __________

14. $-\frac{5}{2}, \frac{15}{8}, -\frac{45}{32}, \ldots$ __________

15. $\frac{2}{3}, \frac{4}{9}, \frac{8}{27}, \ldots$ __________

Find the indicated term for each given geometric sequence.

16. a_6 for the sequence with $a_1 = -2$ and $r = \frac{1}{2}$ ________________

17. a_8 for the sequence with $a_1 = 1.3$ and $r = 2$ ________________

18. a_{10} for the sequence with $a_1 = 0.1$ and $r = 3$ ________________

19. a_{12} for the sequence with $a_1 = -500$ and $r = 0.8$ ________________

20. a_6 for the sequence with $a_1 = -8748$ and $r = -\frac{1}{3}$ ________________

Find the first term of each geometric sequence with the given characteristics.

21. $a_3 = 12.6$ and common ratio $r = 3$ ____________

22. $a_4 = 6\sqrt{2}$ and common ratio $r = \sqrt{2}$ ____________

23. $a_5 = \frac{1}{24}$ and common ratio $r = -\frac{1}{2}$ ____________

24. $a_2 = 1.2$ and $a_5 = 0.0096$ ____________

25. $a_2 = -2$ and $a_4 = -\frac{1}{8}$ ____________

Evaluate the requested sum of each geometric sequence.

26. S_4 for the sequence with $a_1 = 1.4$ and $r = 4$ ____________

27. S_5 for the sequence with $a_1 = 20$ and $r = \frac{1}{10}$ ____________

28. S_7 for the sequence with $a_1 = 5$ and $r = -2$ ____________

29. S_5 for the sequence with $a_1 = 81$ and $r = -\frac{2}{3}$ ____________

30. S_{10} for the sequence $2, 6, 18, \ldots$ ____________

31. S_8 for the sequence $80, 40, 20, \ldots$ ____________

32. S_7 for the sequence $1215, 810, 540, \ldots$ ____________

33. S_5 for the sequence with $a_3 = 28$ and $r = 2$ ____________

34. S_6 for the sequence with $a_4 = -1$ and $r = -\frac{1}{3}$ ____________

35. S_4 for the sequence with $a_4 = \frac{5}{2}$ and $r = \frac{2}{5}$ ____________

36. S_5 for the sequence with $a_5 = 144$ and $a_3 = 36$ ____________

37. S_6 for the sequence with $a_5 = \frac{1}{16}$ and $a_2 = -\frac{1}{2}$ ____________

38. S_∞ for the sequence with $a_1 = -5$ and $r = \frac{2}{5}$ ____________

39. S_∞ for the sequence with $a_1 = 3$ and $r = \frac{4}{3}$ ____________

40. S_∞ for the sequence with $a_1 = 270$ and $r = \frac{3}{4}$ ____________

Express each repeating decimal as a fraction by writing it as an infinite geometric sum and using the formula for S_∞.

41. $0.5555\ldots = 0.\overline{5}$ ____________

42. $0.0404\ldots = 0.\overline{04}$ ____________

43. $1.2222\ldots = 1.\overline{2}$ ____________

44. $4.16666\ldots = 4.1\overline{6}$ ____________

45. $3.0484848\ldots = 3.0\overline{48}$ ____________

Find the total amount accumulated in an annuity under each of the following conditions.

46. Annual deposit: $2500; interest rate: 3.5%; after 20 years.

____

47. Annual deposit: $1600; interest rate: 2.6%; after 18 years.

____

48. Annual deposit: $3250; interest rate: 4.1%; after 25 years.

____

How much should be invested per year in an annuity that pays 2.5% annually if you would like to accumulate each amount below?

49. $25,000 after 20 years

____

50. $46,000 after 25 years

____

Chapter 9 Section 4: Mathematical Induction and the Binomial Theorem

Practice Problems

Use a separate sheet of paper to write the following proofs.

1. Use mathematical induction to prove that the following identity:
$$2+4+6+8+\ldots+2n=n(n+1)$$
is true for all positive integers n.

2. Use mathematical induction to prove that the following identity:
$$4+8+12+\ldots+4n=2n(n+1)$$
is true for all positive integers n.

3. Use mathematical induction to prove that the following identity:
$$3+6+9+\ldots+3n=\frac{3n(n+1)}{2}$$
is true for all positive integers n.

4. Use mathematical induction to prove that the following identity:
$$1+3+5+\ldots+(2n-1)=n^2$$
is true for all positive integers n.

5. Use mathematical induction to prove that the following identity:
$$1+2+2^2+\ldots+2^{n-1}=2^n-1$$
is true for all positive integers n.

6. Use mathematical induction to prove that the following identity:
$$1\cdot2+2\cdot3+3\cdot4+\ldots+n(n+1)=\frac{n(n+1)(n+2)}{3}$$
is true for all positive integers n.

7. Use mathematical induction to prove that the following identity:
$$1\cdot3+2\cdot4+3\cdot5+\ldots+n(n+2)=\frac{n(n+1)(2n+7)}{6}$$
is true for all positive integers n.

8. Use mathematical induction to prove that the expression:
$$n^3-n+3$$
is divisible by 3 for all positive integer n.

9. Use mathematical induction to prove that the expression:
$$7^n-4^n$$
is divisible by 3 for all positive integer n.

10. Use mathematical induction to prove that the expression:
$$8^n-3^n$$
is divisible by 5 for all positive integer n.

11. Use mathematical induction to prove that the expression:

$$2^{3n} - 1$$

is divisible by 7 for all positive integer n.

12. Use mathematical induction to prove that the following relationship:

$$n < 3^n$$

is true for all positive integer n.

13. Use mathematical induction to prove that the following relationship:

$$2^n < 2^{n+1}$$

is true for all positive integer n.

14. Use mathematical induction to prove that the following relationship:

$$2n \leq 2^n$$

is true for all positive integer n.

Use the coefficients of Pascal's triangle to expand each binomial below.

15. $(3+2x)^5$

____

16. $(4-x)^4$

____

17. $(a+b)^6$

____

18. $(q-p)^6$

____

Evaluate each expression containing factorials.

19. $\frac{7!}{4!3!}$ ______________________________

20. $\frac{6!2!}{3!3!}$ ______________________________

21. $\frac{5!3!}{4!4!}$ ______________________________

22. $\frac{10!}{6!3!}$ ______________________________

Use the binomial expansion to find the specified term for each indicated binomial.

23. 6th term of the binomial expansion for $(2-3x)^7$ ____________________

24. 3rd term of the binomial expansion for $(a+2x)^8$ ____________________

25. 8th term of the binomial expansion for $(3x-2)^8$ ____________________

26. 7th term of the binomial expansion for $(ax-3)^7$ ____________________

27. 5th term of the binomial expansion for $(b-ax)^6$ ____________________

CPSIA information can be obtained
at www.ICGtesting.com
Printed in the USA
LVHW01s1904310518
579152LV00005B/8/P

9 781609 278830